P9-ARK-624

MODERN ASPECTS OF REFLECTANCE SPECTROSCOPY

MODERN ASPECTS OF REFLECTANCE SPECTROSCOPY

Edited by
WESLEY W. WENDLANDT
Department of Chemistry
University of Houston
Houston, Texas

Proceedings of the American Chemical Society Symposium on Reflectance Spectroscopy, held September 11-12, 1967, in Chicago, Illinois

PLENUM PRESS • NEW YORK • 1968

Library of Congress Catalog Card Number 68-19188

A Division of Plenum Publishing Corporation
227 West 17 Street, New York, N. Y. 10011

Printed in the United States of America

This volume is dedicated to all workers
in the field of reflectance spectroscopy

FOREWORD

This volume contains all of the papers presented at the American Chemical Society Symposium on Reflectance Spectroscopy. The Symposium was presented under the sponsorship of the Division of Analytical Chemistry, and was held on September 11 and 12, 1967, at the 154th National Meeting of the American Chemical Society, Chicago, Illinois.

The papers presented herein represent a renaissance of interest in reflectance spectroscopy. The technique of reflectance spectroscopy is not, of course, a new technique, however, it has only been applied to problems of a chemical interest in the last decade or so. The instrumentation for this technique in the ultraviolet, visible, and near infrared regions of the spectrum has been available for many years. New and exciting research is being carried out at the present time to extend these techniques to the infrared and far infrared regions as well.

It is a pleasure for the Editor to express his gratitude to Drs. John K. Taylor and E. C. Dunlop of the Division of Analytical Chemistry, ACS, for their cooperation in making the Symposium a reality. The assistance of Miss Julie Norris of the University of Houston for her typing and manuscript organization skill is greatly appreciated. And lastly, but certainly not the least, the Editor would like to acknowledge the cooperation of all of the contributors to this volume. Certainly without their cooperation, this Symposium would not have been a success.

Wesley W. Wendlandt

November 1, 1967
Houston, Texas 77004

CONTENTS

THE PRESENT STATUS OF DIFFUSE REFLECTANCE THEORY*

Harry G. Hecht

University of California, Los Alamos Scientific

Laboratory, Los Alamos, New Mexico 87544

ABSTRACT

A review of progress in the theoretical interpretation of diffuse reflectance spectra is given. The Kubelka-Munk solution is discussed in terms of the more general theory of radiative transfer, and the layer and particle models of Johnson and Melamed are also considered. The nature of multiple scattering processes, reflectance of non-homogeneous materials, and methods of chemical analysis are specialized topics which are discussed in some detail.

INTRODUCTION

The radiation reflected by diffusing media can be regarded as a superposition of two distinct components. The first is due to specular, or regular, reflection at the surface of the medium, for which the well-known Fresnel equations apply. For qualitative discussion, it will be sufficient to consider the simple case of perpendicular incidence, for which the specular reflectivity is given by

$$r = [(n-1)^2 + \kappa_o{}^2] / [(n+1)^2 + \kappa_o{}^2] \qquad (1)$$

where n is the relative refractive index of the sample material and κ_o is the absorption index defined through the Bouger, or Lambert law,

$$I = I_o \exp[-4\pi \kappa_o d/\lambda] = I_o \exp[-\alpha d] \qquad (2)$$

Here d is the distance traveled in the absorbing material for which

* Work done under the auspices of the U. S. Atomic Energy Commission.

the radiation is reduced from I_o to I, and $\alpha = 4 \pi \kappa_o/\lambda$ is the absorption coefficient characteristic of the substance.

The second component of reflection comes about because the radiation which penetrates the surface of the medium undergoes multiple scattering at the surfaces of the individual particles. A portion of the radiation is returned to the surface by the scattering process to emerge as the diffuse reflection. During their excursions into the sample layer, at least some parts of the diffused radiation penetrate sample particles, so that this component also contains valuable absorption information. The actual attenuation with distance follows a relation of the same form as Eq. (2), except that α should be interpreted as a mean absorption coefficient for the particulate sample, and d becomes the mean penetrated layer thickness.

Either the specular or the diffuse component of reflection could be used to determine the absorbing properties of a given medium. However, most real diffusers reflect a composite of the two in an unknown proportion. This is further complicated by the fact that absorption affects the two components differently. From Eq. (1), it will be observed that as the absorption becomes stronger, the specular reflectivity increases. On the other hand, the diffuse radiation penetrates the sample to a certain extent and becomes exponentially attenuated (see Eq. (2)), so that an increase in the absorbance of the material decreases the diffuse component. Thus, the two types of reflection complement each other in such a way that the reflection spectrum in general shows very much less structure than the transmission spectrum when both components are present. For this reason, one tries to minimize either one or the other of these components for spectrophotometric work. For the applications with which we will be concerned, the diffuse component will be of primary interest, so we will wish to minimize the effects due to specular reflection. The ways in which this can be done will be brought out during the course of the following discussion of some experimental results, which will also make us aware of some of the features of reflectance spectra which must be accounted for by an acceptable theory.

SOME FEATURES OF REFLECTANCE SPECTRA

Figure 1 shows the reflectance (recorded in this case as $\log(R_{standard}/R_{sample})$, where $BaSO_4$ is the standard) of a strong absorber $KMnO_4$, as a function of particle size for both the pure substance and for $K(Mn,Cl)O_4$ mixed crystals containing 0.17 mole-% $KMnO_4$. These results are taken from the work of Kortüm and Schottler.[1] There are three features in particular which are worthy of note: a) The effect of particle size is much more pronounced in the case of the mixed crystals. b) The observed

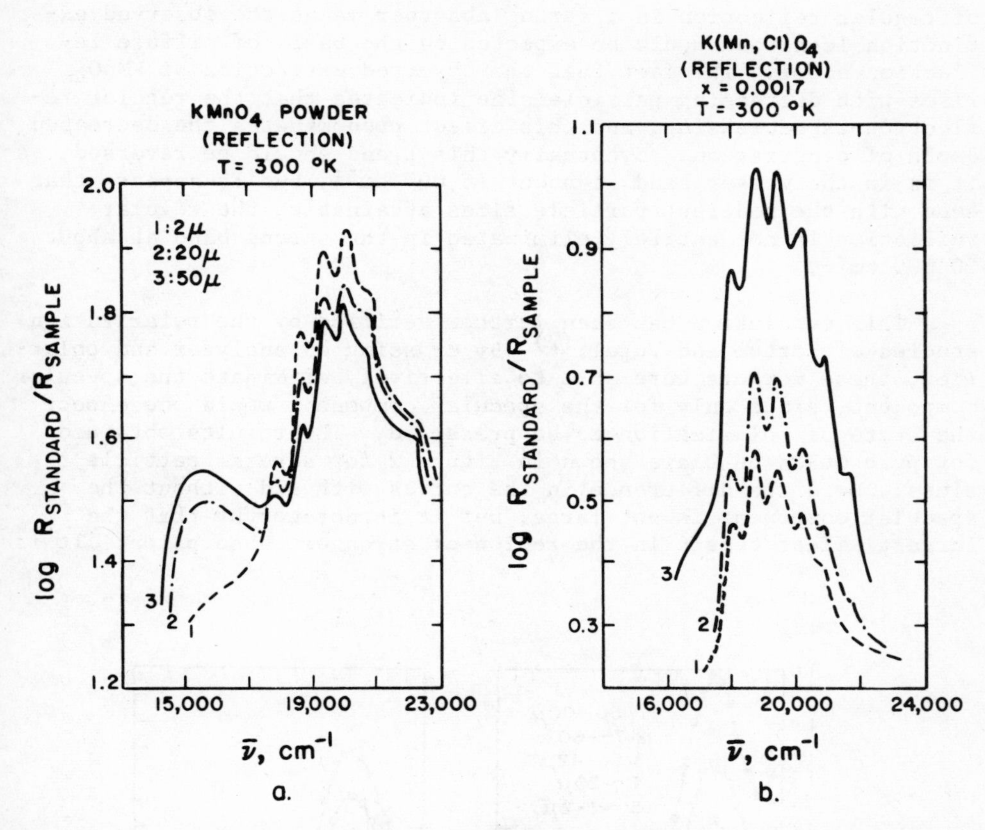

Figure 1. Reflectance curves of (a) pure $KMnO_4$ and (b) $K(Mn,Cl)O_4$ mixed crystals as a function of particle size.[1]

extinctions are only about one unit larger in the pure $KMnO_4$ than in the $K(Mn,Cl)O_4$ mixed crystals, although the difference would be expected to be larger on the basis of concentration. c) The observed extinction of the pure $KMnO_4$ rises with a decrease of particle size, while the $K(Mn,Cl)O_4$ mixed crystals show the opposite trend.

These observations can be correlated with the depth of penetration of radiation into the absorbing sample. We know that weak absorbers, such as $CuSO_4 \cdot 5H_2O$, appear lighter as the material becomes more finely divided. This is because the scattering becomes more efficient and the radiation does not penetrate the sample as deeply. In a strong absorber, the depth of penetration is greatly reduced, so that one would expect to see less dependence on particle size and this is what is observed. The increased proportion of regular reflection in a strong absorber makes the observed extinction less than would be expected on the basis of diffuse reflection alone. The fact that the observed extinction of $KMnO_4$ rises with decreasing particle size indicates that the regular reflection is decreasing, and this effect predominates the decreased depth of penetration. Eventually this trend should be reversed, as it is in the weaker band at about 16,000 cm^{-1}, but it appears that even with the smallest particle sizes attainable, the regular reflection is not entirely eliminated in the strong band at about 20,000 cm^{-1}.

This conclusion has been further verified by the polarization studies of Kortüm and Vogel.[2] By crossing an analyzer and polarizer, these workers were able to effectively eliminate the specular component, since only for the specular component would one expect the state of polarization to be preserved. The results obtained for pure $CuSO_4 \cdot 5H_2O$ are shown in Figure 2 for several particle sizes. Here the difference in the curves with and without the specular component is not large, but it is noteworthy that the largest effect occurs in the region of strongest absorption. It is

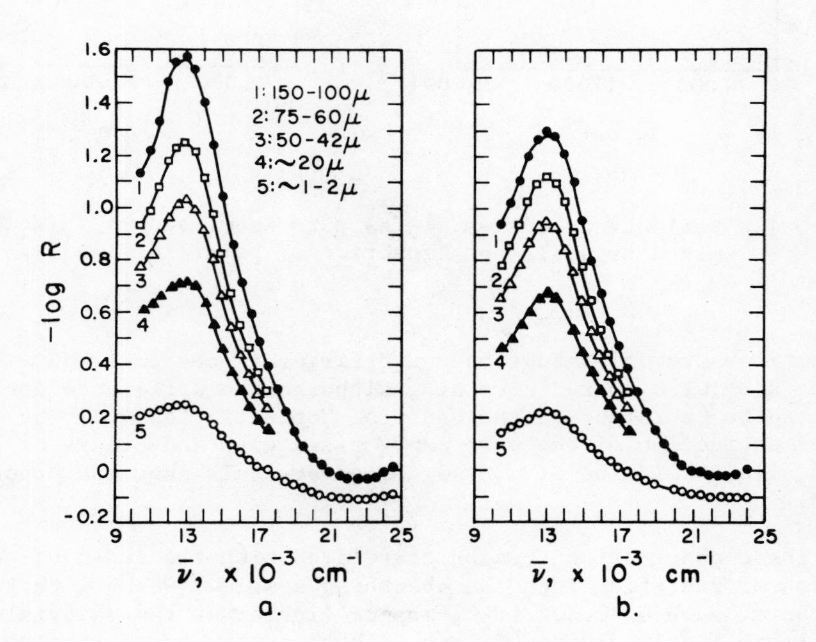

Figure 2. Reflection spectra of pure $CuSO_4 \cdot 5H_2O$ as a function of particle size (a) with polarization films and (b) without polarization films.[2]

also observed that the difference becomes insignificant for the samples with very small particles, so that the specular reflection is virtually eliminated by extensive grinding.

The strong absorber, $KMnO_4$, behaves quite differently. It will be observed by reference to Figure 3 that the extinction is effected very significantly by the polarization films, indicating

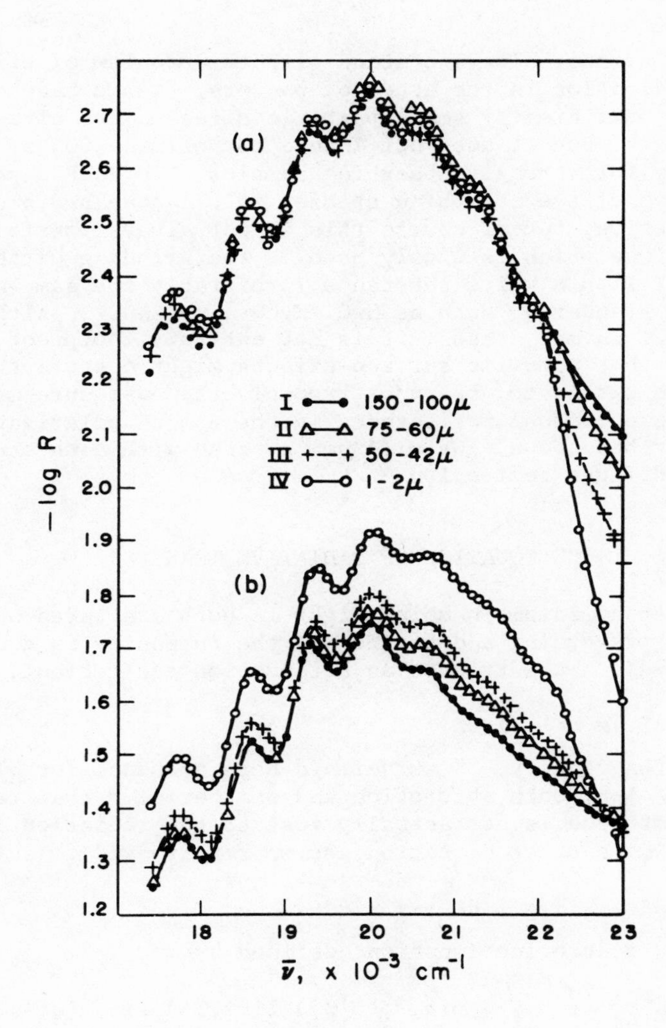

Figure 3. Reflection spectra of pure $KMnO_4$ as a function of particle size (a) with polarization films and (b) without polarization films.[2]

that there is considerable specular reflectivity even for samples composed of very small particles. Note that in this case the elimination of the specular component causes the reflectance to become independent of particle size, in agreement with our supposition that the mean depth of penetration is very small. The diffuse reflectance still arises from scattering processes, however. This is made possible by the statistical distribution of edges and corners of appropriate thickness and orientation near the surface of the sample.

Thus, one must always contend with the problem of eliminating specular reflection in the study of powders. As we have seen, grinding to sufficiently small particle sizes may be effective for weak absorbers, but it does not insure the elimination of specular reflectivity for strongly absorbing samples. In such cases, the incorporation of the absorbing species into a non-absorbing crystal may be effective, but of course this is not always possible. Another technique which is widely used is the grinding of the absorbing material with a white substance (preferably the same as that used as the standard), such as MgO, $MgCO_3$, or $BaSO_4$. Although this is successful in many cases, it is not entirely foolproof since it is possible that specific surface effects might obscure the very features one wishes to observe. Probably the most sure method of observing pure diffuse reflectance is the use of polarization studies as described above. We will now proceed with a theoretical account of diffuse reflection.

THE EQUATION OF RADIATIVE TRANSFER

Consider a medium in which light is both scattered and absorbed. In traversing a distance dS, the intensity is decreased by an amount, -dI. We introduce an attenuation coefficient, κ, such that

$$-dI = \kappa \rho I \, dS \tag{3}$$

where ρ is the density. κ as defined here accounts for <u>all</u> radiation lost to I by both absorption and scattering. That radiation which is scattered is not actually lost to the radiation field, but shows up in some other direction, so we really need to solve

$$-dI = \kappa \rho I \, dS - j \rho \, dS \tag{4}$$

where j is a scattering function, defined by

$$j(\theta,\phi) = \kappa \frac{1}{4\pi} \int_0^{\pi} \int_0^{2\pi} p(\theta,\phi;\, \theta'\phi') \, I(\theta',\phi') \sin\theta' d\theta' d\phi' \tag{5}$$

$p(\theta,\phi;\, \theta'\phi')$ is known as the <u>phase function</u>, which is obviously a measure of the intensity of scatter in the direction (θ,ϕ) of radiation which is incident in the direction (θ',ϕ').

In terms of the optical thickness,

$$\tau = \int_{s'}^{s} \kappa \, \rho \, ds \tag{6}$$

the equation of transfer is

$$(-dI/d\tau) = I - (j/\kappa) \tag{7}$$

It is useful in the solution of problems dealing with plane-parallel layers to let s be the distance along the surface normal, in which case the equation of transfer can be written

$$\mu \frac{dI(\tau,\mu,\phi)}{d\tau} = I(\tau,\mu,\phi) - \frac{1}{4\pi}\int_{-1}^{+1}\int_{0}^{2\pi} p(\mu,\phi;\mu'\phi')\, I(\tau,\mu',\phi')\, d\mu' d\phi' \tag{8}$$

where $\mu = \cos\theta$ and θ defines the angle with respect to the inward surface normal.

THE PHASE FUNCTION

If the direction of scatter is inclined to the direction of incidence by the angle Θ, then

$$p(\cos\Theta)\frac{d\omega}{4\pi} \tag{9}$$

gives the fraction scattered into the solid angle $d\omega$, and the total scatter in all directions is

$$\int_{0}^{4\pi} p(\cos\Theta)\frac{d\omega}{4\pi} \tag{10}$$

If only scattering takes place this integral is unity, but in the general case we can write

$$\int_{0}^{4\pi} p(\cos\Theta)\frac{d\omega}{4\pi} = \omega_o \leq 1 \tag{11}$$

ω_o, the albedo for single scatter, represents the fraction of the light lost by scattering, while the fraction $(1 - \omega_o)$ is lost by other processes, i.e., transformed into other forms of energy.

ω_o can be expressed in terms of the scattering and absorption coefficients, σ and α, as

$$\omega_o = \frac{\sigma}{\sigma + \alpha} \tag{12}$$

and in the simple isotropic case the phase function is

$$p(\cos \Theta) = \omega_o = \frac{\sigma}{\sigma + \alpha} \tag{13}$$

For non-isotropic scatter, the phase function

$$p(\cos \Theta) = \omega_o(1 + x \cos \Theta) \qquad (0 \leq x \leq 1) \tag{14}$$

is frequently used. To be completely general, $p(\cos \Theta)$ can be expanded as a series of Legendre polynomials,

$$p(\cos \Theta) = \sum_{\ell=0}^{\infty} \omega_\ell \, P_\ell(\cos \Theta) \tag{15}$$

but terms of higher order than first degree are ineffectual to a good degree of approximation,[3,4] so that Eq. (14) can be regarded as a rather general phase function.

A SIMPLIFIED SOLUTION FOR THE CASE OF ISOTROPIC SCATTER

Let us consider the case of isotropic scatter in a plane-parallel medium. The result should be independent of the azimuthal angle and Eq. (8) reduces to

$$\mu \frac{dI(\tau,\mu)}{d\tau} = I(\tau,\mu) - \frac{1}{2}\omega_o \int_{-1}^{+1} I(\tau,\mu')d\mu' \tag{16}$$

It will be assumed that the radiation can be divided into two parts; that which travels toward the surface we will call J, and that which travels away from the surface we will call I. With this simplification, Eq. (16) can be separated into a pair of equations,

$$\begin{aligned} \frac{1}{2}(dI/d\tau) &= I - (\omega_o/2)(I + J) \\ -\frac{1}{2}(dJ/d\tau) &= J - (\omega_o/2)(I + J) \end{aligned} \tag{17}$$

where the factor $\pm \frac{1}{2}$ on the left hand side arises from an average of μ over all possible angles relative to the surface normal (see ref. (5), pg. 58). Defining $s = \sigma/(\sigma + \alpha)$ and $k = 2\alpha/(\sigma + \alpha)$, these coupled differential equations can be written,

$$(dI/d\tau) = -(k + s)I + sJ$$
$$(dJ/d\tau) = (k + s)J - sI \qquad (18)$$

Equations (18) are of the form introduced by Schuster,[6] Schwarzschild,[7] and Kubelka and Munk,[8] and the solution can readily be shown to be

$$I = A(1 - \beta)e^{\kappa\tau} + B(1 + \beta)e^{-\kappa\tau}$$
$$J = A(1 + \beta)e^{\kappa\tau} + B(1 - \beta)e^{-\kappa\tau} \qquad (19)$$

where

$$\kappa = \sqrt{k(k + 2s)} \quad \text{and} \quad \beta = \sqrt{k/(k + 2s)} \qquad (20)$$

Evaluation of A and B for the boundary conditions,

$$I = I_o \quad \text{at } \tau = 0$$
$$I = J = 0 \quad \text{at } \tau = \infty \qquad (21)$$

leads to

$$R = J_{\tau=0}/I_o = (1 - \beta)/(1 + \beta) \qquad (22)$$

or

$$f(R) = (1 - R)^2/2R = k/s \qquad (23)$$

f(R) is known as the remission function and it has been widely used for the interpretation of reflectance data. Values of this function have been tabulated for convenience.[5,9]

EXACT SOLUTIONS FOR MATRICES WHOSE REFRACTIVE INDEX IS UNITY

Rigorous solutions to the equation of radiative transfer can be developed in the case where the refractive index of the medium containing the scattering particles is unity. The technique stems from the success of the approximate solution discussed in the previous section, in which the radiation field was divided into the two components along the surface normal. A generalization of this procedure involves the solution of Eq. (16) for an arbitrary number of directions, the technique for which has been developed by Chandrasekhar.[10]

The integrodifferential equation of radiative transfer can be simplified by expressing the integral in terms of a Gaussian quadrature, the number of integration points determining the number of directions for which a solution is sought. Equation (16) thus reduces to the set of linear differential equations

$$\mu_i \, (dI_i/d\tau) = I_i - \tfrac{1}{2}\, \omega_o \sum_{j=-n}^{+n} a_j I_j \tag{24}$$

where the a_j are the Gaussian weighting factors given in the usual way by

$$a_j = [1/P_n'(\mu_j)] \int_{-1}^{+1} [P_n(\mu)/(\mu - \mu_j)] \, d\mu \tag{25}$$

where μ_j is one of the zeros of the Legendre polynomial, $P_n(\mu)$.

Exact solutions are obtained by following this line of reasoning and passing to the limit, $n = \infty$. It is found that the solutions can be expressed in terms of the so-called H-integrals,

$$H(\mu) = 1 + \mu \, H(\mu) \int_0^1 [\Psi\,(\mu')/(\mu + \mu')] \, H(\mu') \, d\mu' \tag{26}$$

and the various moments of $H(\mu)$ defined by

$$a_n = \int_0^1 \mu^n \, H(\mu) \, d\mu \tag{27}$$

Here Ψ is the <u>characteristic function</u>, being equal to $\frac{1}{2}\,\omega_o$ in the isotropic case, and $(1/2)\omega_o[1 + x(1 - \omega_o)\mu^2]$ for the phase function $\omega_o(1 + x \cos \Theta)$. Tables of the H- and α-integrals have been given by Chandrasekhar (ref. (10), pgs. 125, 139, 141, 328).

Since the development of the solutions is rather tedious, we will merely quote the results as given by Giovanelli[4]: The total reflectance for light incident in the direction μ_o for isotropic scatter is

$$R(\mu_o) = 1 - H(\mu_o)\,(1 - \omega_o)^{1/2} \tag{28}$$

and for scatter by the phase function $\omega_o(1 + x \cos \Theta)$ is

$$R(\mu_o) = 1 - H(\mu_o) \left\{1 - [(\omega_o/2)(\alpha_o - c\alpha_1)]\right\} \tag{29}$$

where

$$c = x(1 - \omega_o)\, \omega_o \, [\alpha_1/(2 - \omega_o\alpha_o)] \tag{30}$$

For diffused incident radiation the result is

$$R_D = 1 - 2(1 - \omega_o)^{1/2} \alpha_1 \tag{31}$$

for isotropic scatter, and

$$R_D = 1 - 2\alpha_1 \left\{1 - [(\omega_o/2)(\alpha_o - c\alpha_1)]\right\} \tag{32}$$

for scatter according to $\omega_o(1 + x \cos \Theta)$.

SOLUTIONS FOR MATRICES WHOSE REFRACTIVE INDEX IS GREATER THAN UNITY

A problem of considerable practical importance is the solution of the equation of radiative transfer for scattering particles imbedded in a medium whose refractive index exceeds unity. These conditions apply, for example, when considering the reflecting properties of paints and turbid solutions of various sorts. The reflectance in this case depends not only upon the nature of the surface of the medium, but also upon the refractive index, because of the possibility of internal reflections.

The surface in these cases is best described as specular reflecting, but the solution for a diffusing surface is also of interest, since it provides the opposite limit defining the range within which all real media lie. Solutions have been given by Giovanelli[4] for isotropic scatter in a medium with a perfectly diffusing surface, and for isotropic scatter or scatter according to the phase function $\omega_o(1 + x \cos \Theta)$ in a medium with a specular reflecting surface. Reflectance curves can be constructed from tables, contained in Giovanelli's paper, which apply to any of a large number of experimental conditions.

TEST OF THE THEORETICAL RESULTS

Before making a comparison with experimental data, it is instructive to see how closely the Kubelka-Munk solution (Eq. (23)) approximates that obtained by exact methods. For this purpose we consider the simple case of diffused incident radiation and ideal isotropic scatter within the powder medium, which conditions are inherent in the usual Kubelka-Munk derivation. The exact solution (Eq. (31)) gives the solid curve of Figure 4, while the dashed curve represents the Kubelka-Munk solution. It will be observed that the two curves are very similar, differing at most by just a few percent. Blevin and Brown[11] have shown that by normalizing the Kubelka-Munk curve to coincide with the exact solution at $R = 0.5$, a factor of no practical significance for the usual

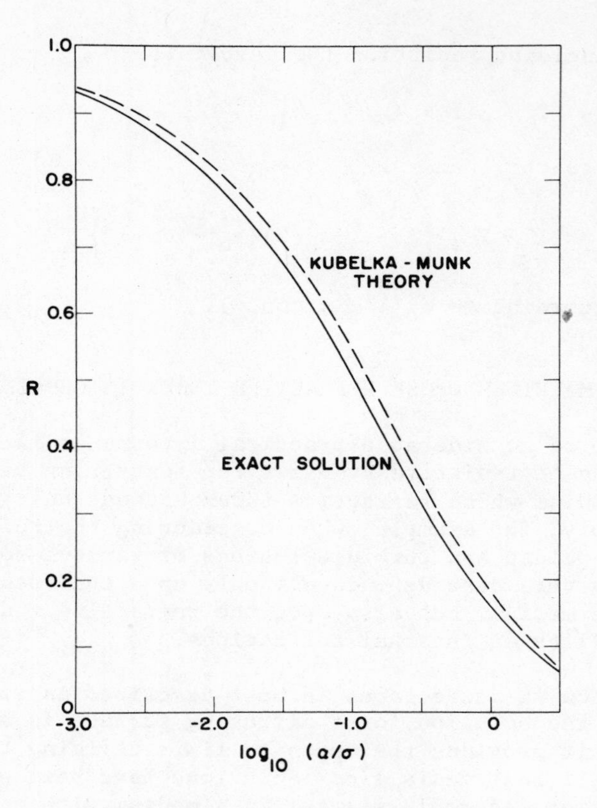

Figure 4. Exact solution to the equation of radiative transfer for diffused incident radiation and isotropic scatter (solid curve), compared with the Kubelka-Munk solution (dashed curve).

comparative type applications*, the maximum deviation is only about 0.01 unit of reflectance. Errors of about this magnitude are introduced by the assumption of isotropic scatter, which never actually occurs (see next section).

Thus, in those cases where the above-mentioned conditions are fulfilled, the remission function can be regarded as an excellent approximation of the exact solution. Of course, we have a direct comparison only in this simple case, but there is good reason to believe that the conditions assumed here are closely approximated in many cases where precautions are taken to eliminate the specular component, as previously mentioned. Even where the scattering is

* Recently, considerable interest has been shown in absolute reflectance measurements.(12,13) Most chemical applications involve identification or determination of concentration which depend on the relative reflectance.

known to be angular dependent, the solution for the isotropic case is of interest, since it can be shown that the isotropic solution is a good approximation if the actual scattering coefficient is replaced by an effective scattering coefficient, σ_e, where

$$\sigma_e = \sigma(1 - \bar{\mu}) \tag{33}$$

and

$$\bar{\mu} = \int_{-1}^{+1} I(\mu)\mu d\mu \Big/ \int_{-1}^{+1} I(\mu) d\mu \tag{34}$$

with μ defined as the cosine of the angle between the directions of incidence and scatter, and $I(\mu)$ is the scattered intensity in the direction μ.[14,15]

As a typical example of the validity of this reflectance theory, the reflectance and transmission data for a didymium filter glass, taken from the work of Kortüm, Braun, and Herzog,[16] are shown in Figure 5. Here the transmission curve is seen to

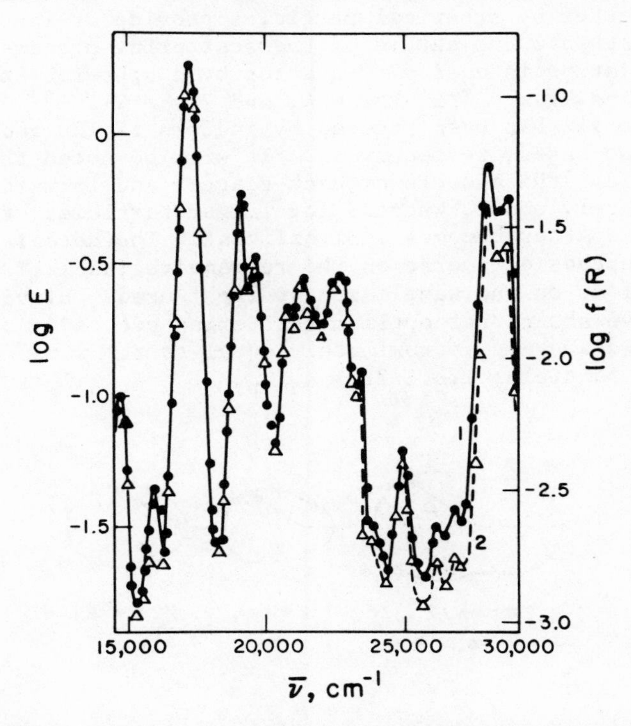

Figure 5. Transmittance spectrum (curve 1) and reflectance spectrum (curve 2) of a didymium-glass filter. The transmittance spectrum was measured relative to quartz glass 0.5 mm thick, and the reflectance spectrum relative to powdered colorless glass.[16]

superimpose the reflectance curve very well, the latter having been determined for the same glass sample after it had been ground to a small particle size. It is log f(R) which is plotted here, so that the actual reflectance curve is subject to a displacement by the constant, log s. Such a curve is known as a characteristic color curve. In such work the remission function is of course expressed in terms of the reflectance relative to a white standard, which, in this case, was a powdered colorless glass.

A systematic deviation of the curves is readily apparent in the short wavelength region of the spectrum. This difference is attributed to a change in the scattering coefficient, the nature of which we will now consider.

THE SCATTERING COEFFICIENT

No general theory of multiple scattering in dense media has been worked out as yet, but the extensive investigations of Mie[17] of single scatter by spherical particles provide at least a qualitative insight into the nature of the scattering process. Figure 6 shows the relative intensity of scatter by a spherical particle for the three cases, $2\pi r = \lambda/2$, $2\pi r = \lambda$, and $2\pi r = 2\lambda$.[14] The scattering indicatrix has been reduced by factors of 100 and 5000 in the latter two cases, respectively. It will be noted that for small particles, the amounts of back scatter and forward scatter are approximately equal, whereas for larger particles, the proportion of back scatter becomes insignificant. The actual indicatrix of scatter depends of course on the refractive index of the particles, as well as on the wavelength of light used. Blevin and Brown[15] have shown that optimum scattering occurs in dense media for particle diameters approximately equal to the wavelength of light in the particle; i.e., $2r \approx \lambda_o/n_p$.

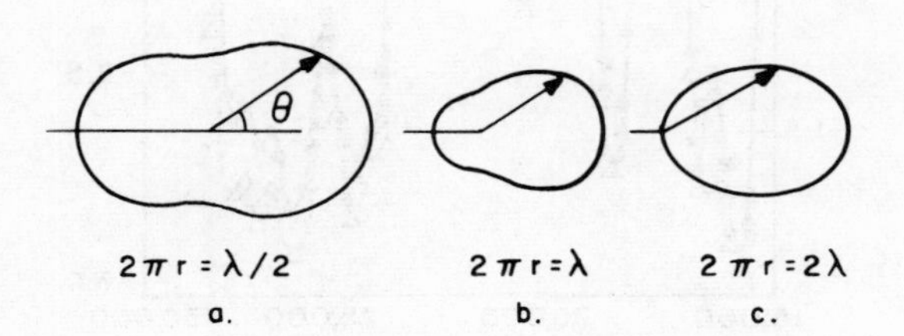

Figure 6. Relative intensity of radiation scattered through the angle θ by spheres of refractive index 1.33 for (a) $2\pi r = \lambda/2$, (b) $2\pi r = \lambda$, and (c) $2\pi r = 2\lambda$. The scales for cases (b) and (c) have been reduced by the factors 100 and 5000, respectively.[14]

It should be noted also by reference to Figure 6 that in _no_ case is the scattering isotropic. In dense media, the situation could be quite different, however. Rozenberg[18] has indicated that close packing of particles can cause important changes in the scattering indicatrix due to the interference of light scattered by neighboring particles, and Blevin and Brown[19] have observed effects at very high particle concentrations which were attributed to this effect. The approximation of isotropic scatter in dense media is adequate for most work. This is borne out by the study of Blevin and Brown[15] in which the reflectance for isotropic scatter was compared with the reflectances for the phase functions $1 + P_1(\mu)$, $1 + P_2(\mu)$, and $1 + P_3(\mu)$, which are shown in Figure 7. It was found that the reflectance curves for all these cases were nearly superimposable.

The mathematical derivations of Mie[17] for single scatter show that for particles whose dimensions are small with respect to the wavelength of radiation, the scattering assumes a λ^{-4}-dependence characteristic of Rayleigh scattering. In the other limit, i.e., where the particle dimensions are large with respect to the wavelength of the light, the scattering approaches a λ-independence. Thus, we might expect to observe a wavelength dependence for

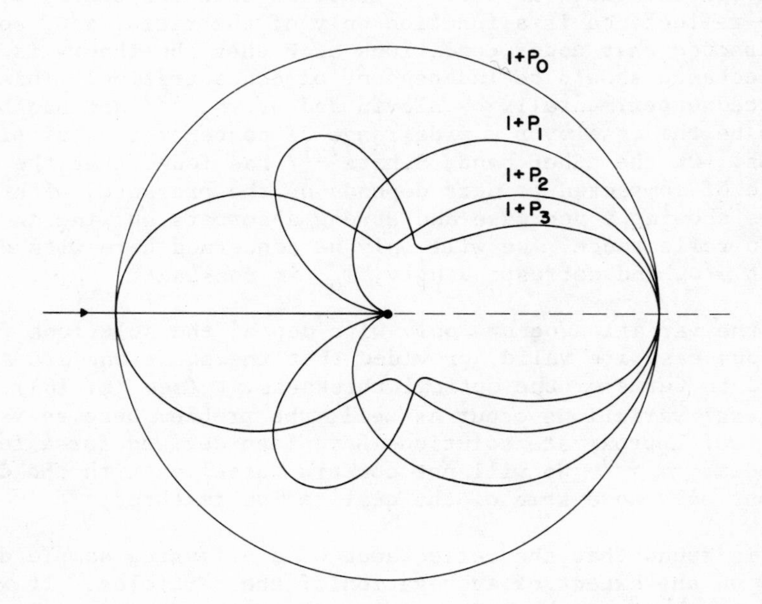

Figure 7. Indicatrix for the scattering phase functions, $1 + P_0(\mu)$, $1 + P_1(\mu)$, $1 + P_2(\mu)$, and $1 + P_3(\mu)$.

finely divided particles, but none for course-grained samples. This effect has been demonstrated for a didymium filter glass by Kortüm, Braun and Herzog.[16] In a more quantitative study, Kortüm and Oelkrug[20] determined the wavelength dependence of scatter to be $\lambda^{-2.6}$ to $\lambda^{-3.6}$ for particles such that $d < \lambda$, $\sim \lambda^{-1}$ for $d \approx \lambda$, and λ^{-1} to λ^{0} for $d < \lambda$, in excellent agreement with the qualitative predictions. These same workers[20] considered the effect of the λ-dependence of scatter on the shape and position of absorption bands determined by reflectance. The effect was found to be rather small except for very broad bands.

THE EFFECT OF INHOMOGENEITIES

A large proportion of the systems whose reflectance is of interest can be classified as non-uniform diffusers. This occurs, for example, with samples where clustering of particles occurs, or in paints with a non-uniform dispersion of pigment. Such cases are naturally more difficult to treat in general, but some progress has been made in understanding the nature of these effects.

As a first approximation, it seems reasonable to suppose that both the scattering and absorption coefficients vary linearly with concentration, since an increase in the density of particles should enhance both effects. It will be recalled that the theory predicts that the reflectance is a function only of the ratio, α/σ, so it is to be expected that under conditions such that the theory is valid, the reflectance should be independent of concentration. This has been tested experimentally by Blevin and Brown,[19] and has been found to be the case over a wide range of concentration of pigment particles. On the other hand, Schatz[21] has found that the reflectance of compacted powders depends on the pressure, with weak absorbers showing a decrease and strong absorbers showing an increase in reflectance. We will only be concerned here with cases such that α/σ, and correspondingly, ω_o, is constant.

If the variation occurs only with depth, the solutions for the homogeneous case are valid, provided that the equations are always expressed in terms of the optical thickness, τ (see Eq. (6)).[22,23] When lateral variations occur as well, the problem becomes very complex, but approximate solutions have been derived for a few model media.[23-26] We will not concern ourselves with the details, but only note some of the qualitative features.

It is found that the reflectance of a diffusing sample depends markedly on the extent of aggregation of the particles. It will be recalled that in uniform media, the reflectance depends only on ω_o, but not on the attenuation, κ. This is not the case with clustered particles. Although ω_o is still constant, the reflectance may typically be lowered by ca. 20 or 30% if large, well-spaced aggregates are present. This is not surprising, for we know that the diffusely reflected component must be multiply scattered by the individual particles, and if aggregation of these occurs, the concentration of scattering centers is reduced and hence the diffuse reflectance is reduced.

It might have been supposed that the reflectance would increase as the separation between aggregates is reduced, but in fact, just the opposite is true. This decrease of reflectance is explained as follows[24]: Even in closely packed media, only about half the volume is taken up by the scattering particles, and hence there are cavities which may be penetrated by the radiation either directly or following reflection from the outer halves of the aggregate particles, where it is largely absorbed. This leads to a low reflectance for compacted aggregates.

Another result of these studies which is of considerable interest, is that the reflectance of an incompletely dispersed mixture of diffusers is not necessarily intermediate, but may be much lower than those of the individual components of which the mixture is composed. This is an effect of inhomogeneity, and not equivalent to the effect observed by Schatz,[27] wherein a different effective reflectance was postulated to account for a disproportionately reduced internal reflectance of the transparent component of some mixtures.

Thus, a constancy of ω_o does not insure reproducible reflectance measurements; considerable care must be taken to ensure the complete dispersion of scattering particles for the accurate determination of optical properties. Such considerations are of particular significance to the paint industry, where one wishes to achieve the maximum hiding power for a given quantity of pigment.

TECHNIQUES OF CHEMICAL ANALYSIS

The remission function given in Eq. (23) can be written

$$F(R) = \text{constant} \times c \tag{35}$$

where c is the molar concentration and it has been assumed that $k = 2.303\,\epsilon\,c$, where ϵ is the molar extinction coefficient. The validity of this relationship has been tested experimentally for $K(Mn,Cl)O_4$ mixed crystals by Kortüm and Schottler[1] and for K_2CrO_4 mixed with MgO by Kortüm and Schreyer.[28] The remission function was found to be linearly related to the concentration, at least over a limited concentration range, so that reflectance studies can be used for quantitative chemical analysis in much the same way that transmission spectrophotometry is used.

The range of applicability of the remission function in this context is rather limited, presumably due to such factors as the influence of the specular component. Because of this, a number of empirical relationships have been proposed to extend the range, among which are log R <u>vs</u>. C^2, R <u>vs</u>. $C^{1/3}$, R <u>vs</u>. $[C/(1 - C)]^{1/2}$,[29] and f(R) <u>vs</u>. k log C.[30]

For the analysis of mixtures, one might assume an additivity for k and s, such that

$$f(R) = \sum_i C_i k_i / \sum_i C_i s_i \tag{36}$$

where C_i is the concentration of the ith component in the mixture. If the concentration of the absorbing materials is small, as is frequently the case, then the scattering is predominated by the basis material and Eq. (36) becomes

$$f(R) = \sum_i C_i (k_i/s) \tag{37}$$

This linear equation can be solved for the concentrations by observing the reflectance at as many wavelengths as there are components in the mixture. Such techniques have been used by Foote,[31] Everhard, Dickcius, and Goodhart,[32] and Frei, Ryan, and Lieu.[33]

A method of analysis has been proposed by Giovanelli[14,34] based upon the observation that the reflectance in the range ca. 0.2 to 0.7 is approximately linear in log (α/σ) (see Figure 4). We let α_b be the absorption coefficient of the background, determined in the first approximation by graphical interpolation between those points where α_a, the absorption coefficient of the absorbing component, is known to be small. In this way we can approximate ΔR_o, the decrease in reflectance at the band center, and W, the half-width of the absorption band, by assuming a usual Lorentzian form,

$$\alpha_a = CA / [1 + (\Delta\lambda/W)^2] \tag{38}$$

where A is a constant characteristic of the absorption band and C is the concentration. If ΔR_o is small, it can be shown that the decrease in reflectance anywhere in the band is simply

$$\Delta R/\Delta R_o = 1/[1 + (\Delta\lambda/W)^2] \tag{39}$$

so one can readily construct the reflectance curve for a given set of parameters and see if the background absorption, α_b, has been well chosen. If not, an improved estimate can be made and the cycle repeated to self consistency. If there are overlapping absorption curves, the procedure becomes more complicated, but is handled in the same, straight-forward way.

Once ΔR_0 has been determined, Figure 4 can be used to determine α_a/σ. Of course it is only this ratio which is determined, so that this method, in common with the others previously mentioned, only yields relative concentrations but this is not a serious limitation for most purposes.

If we have two different samples, characterized by $\alpha_{a1}/\sigma_1 = X$ and $\alpha_{a2}/\sigma_2 = Y$, their relative concentrations can be determined as follows: We mix equal parts of the two samples by volume, and the reflectance of the resulting mixture is determined by α_a/σ, where $\alpha_a = \frac{1}{2}(\alpha_{a1} + \alpha_{a2})$ and $\sigma = \frac{1}{2}(\sigma_1 + \sigma_2)$. If we let $Z = \alpha_a/\sigma$, then

$$\alpha_{a1}/\alpha_{a2} = (Z/Y - 1) / (1 - Z/X) \tag{40}$$

The Giovanelli method has been tested experimentally by Rickard, Moss, and Roper,[35] who found that the cytochrome concentration of yeast samples could be accurately determined for several different line shape functions.

PARTICLE MODELS

The foregoing discussion provides a theoretical basis which is adequate for most chemical applications, the one significant limitation being that no absolute absorption coefficients can be determined. This is because the absorption and scattering characteristics of the medium are phenomenologically accounted for in the two constants, α and σ, with no reference to the nature of the particles composing the sample. Since spectrophotometric analyses usually employ comparative measurements, a great many studies can be carried out in this way, but a completely general theory which allows absolute values for the optical constants to be determined must be based on the properties of the individual sample particles.

Any real sample is composed of particles of various sizes and shapes, and so some idealized model must be conceived before meaningful calculations can be performed. Thus, calculations of reflectance on an individual particle basis are of necessity tied to a specific model of the medium, the nature of which determines the relationship between the microscopic particle characteristics and the reflectance of the sample.

A model which has often been used treats the sample as a set of plane parallel layers, the thickness of each being equal to the mean diameter of the particles, d (see Figure 8). Johnson[36] has treated this problem in some detail, and arrived at the equation,

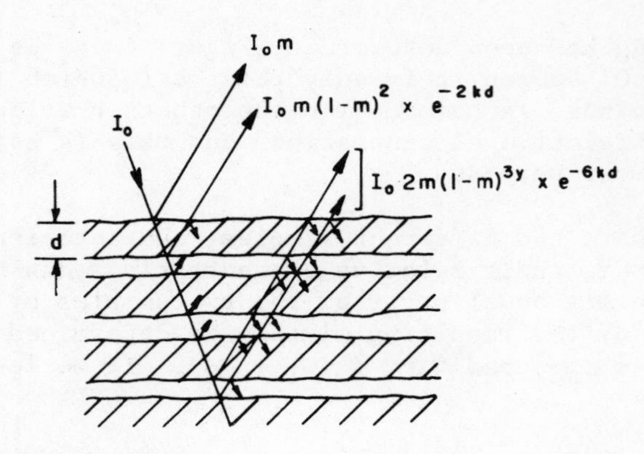

Figure 8. Layer model used by Johnson[36] for the calculation of absolute absorption coefficients from diffuse reflectance.

$$R = m\left[\frac{2(1 - m)^y e^{-2kd}}{2kd - y \ln(1 - m)} + 1\right] \tag{41}$$

by summing the reflections from each of the layers. Here m is the fraction of the incident radiation reflected from the first layer, and y is an adjustable parameter which accounts in part for multiple reflections, which are not considered in the above summation, and for losses due to scattering. Solving for y, subject to the condition that $R = 1$ for $k = 0$, it is found that $y = 2$ for refractive indices less than 1.5. It is interesting to note that k and d enter Eq. (41) only as the product, kd, so that identical reflectances are predicted for any k if appropriate adjustment is made in the particle size, d.

The applicability of Eq. (41) is limited on the one hand by insensitivity to changes in k for large values of kd, and on the other by the rather arbitrary account of multiple reflections for small values of k. Another assumption in Johnson's treatment is that $m = 1.5\ r$, where r is the Fresnel reflectivity given by Eq. (1). This is an attempt to account for the random orientation of particle surfaces with respect to the incident beam, and corresponds to an average incidence of approximately 30°.

In spite of these features, the Johnson model gives a reasonably good account of reflectance for values lying in the range ca. 0.1 to 0.8, although Companion[37] has found some rather large discrepancies in the study of V_2O_5. In addition to the above-mentioned factors, index of refraction anomalies at the absorption bands and the distribution of particle sizes about the mean may also limit the accuracy of this analysis.

A method which circumvents several of the shortcomings of the Johnson model has been given by Melamed,[38] who not only carried out a summation over individual particle reflections, but included all multiple reflections as well (see Figure 9). The final result is of the form (note a typographical error in the original paper),

$$R = 2x\bar{m}_e + \frac{x(1 - 2x\bar{m}_e)T(1 - \bar{m}_e R)}{(1 - \bar{m}_e R) - (1 - x)(1 - \bar{m}_e)TR} \tag{42}$$

where T is the particle transmission, x is the probability that the light emerging from a particle is scattered toward a particle closer to the surface, and $\bar{m}_e$ is the average value of the reflection coefficient of the surface of an individual particle for externally incident radiation.

A specific size and shape for the scattering particles must be assumed in any detailed application of Eq. (42). Melamed assumed that the irregular particles of the diffusing sample can be represented by uniform spheres of the appropriate diameter. Assuming that each surface element of the particle scatters according to the Lambert cosine law, and that $\bar{m}_e$ is determined by an average over

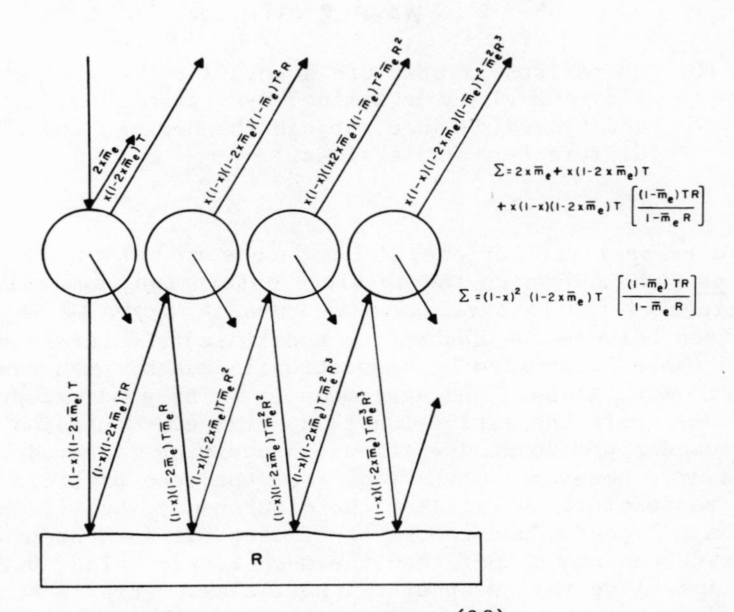

Figure 9. Particle model used by Melamed[38] for the calculation of absolute absorption coefficients from diffuse reflectance.

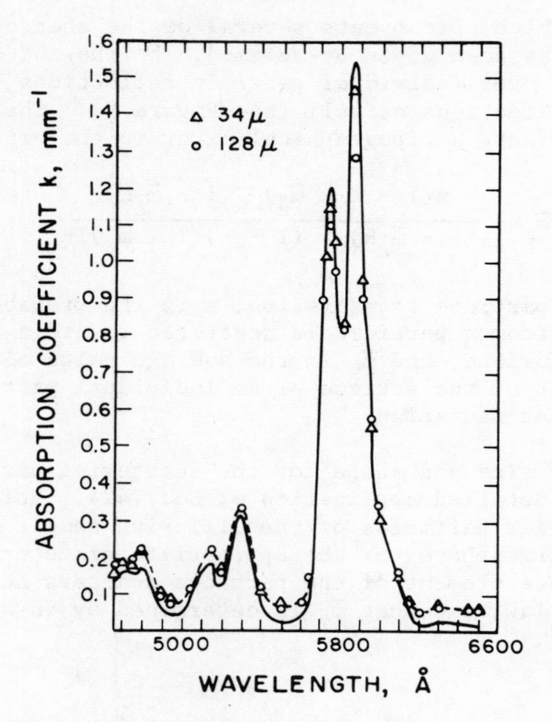

Figure 10. Comparison of absolute absorption coefficients for a didymium glass determined by transmission (solid curve) and by reflectance through the Melamed model for two different particle sizes.[38]

all the Fresnel reflections, Melamed constructed curves of reflectance vs. kd from which the absolute absorption coefficient can be determined if the particle size is known. Figure 10 shows a comparison between the absorption coefficients determined in this way and those determined by conventional transmission techniques for a didymium glass. The agreement is quite good except for small deviations where the extinction is not large. Companion has here again found significant deviations in applying this model to V_2O_5 samples,[37] however. Deviations from the mean particle diameter may be responsible in part for the discrepancy, but it is significant that Companion had to use the factor 0.1 for the probability of upward scatter, rather than the much larger value, 0.284, appropriate for close-packed spherical particles. This is strongly suggestive that the particles actually do pack more compactly than the spherical model would suggest, which does not seem at all improbable since the small crystallites, although irregular in shape, are bounded by planar surfaces which allow for intimate contact.

CONCLUSIONS

The laws governing specular reflectivity have long been well understood, and, from the foregoing discussion, it appears that it can be successfully argued that our knowledge of diffuse reflectance is also quite complete. Through the work of Chandrasekhar[10] and Giovanelli,[4] exact solutions and good approximate solutions to the equation of radiative transfer are available for several cases of interest. It has been shown that the remission function derived on the basis of the simple Kubelka-Munk theory gives an account of diffuse reflectance no more in error than the deviations in the exact solutions encountered by the assumption of different scattering phase functions. Since the remission function represents a convenient analytic approximation to the exact solution which is applicable in many cases, it seems that its continued use for the interpretation of reflectance data can be justified.

The work of Melamed[38] has similarly brought the status of calculations based upon a single particle model into quite a satisfactory state, and allows absolute absorption coefficients to be determined by reflectance measurements. It might be of interest to investigate the effect of other particle shapes, and the effect on the reflectance of averaging over a statistical distribution of particle sizes, but the model has proven sufficiently successful to indicate that these are merely refinements and that the features of reflectance spectra can be accounted for quite successfully on the basis of remarkably simple models.

It seems that most of the difficulty encountered in reflectance work is not with the theory, but with the application. As pointed out in the introductory section, either diffuse or specular reflection can be used for the determination of optical constants, but that we frequently encounter situations where the two are combined in an unknown proportion. The presence of a specular component in any diffusing sample is unavoidable in practice. Even with relatively weak absorbers the difficulty is often encountered, perhaps in some cases due to the fact that with virtually any sample packing technique, the sample particles near the surface are probably preferentially oriented with their largest planar surfaces parallel to the sample surface. The only way of unambiguously avoiding this difficulty is to use polarizers to eliminate the specular component, as Kortüm and Vogel[2] have done.

The particle theories give a more nearly _a priori_ theory of reflectance. However, the reflectance is dependent on the refractive index through the specular reflectivity of the individual particles (see Eq. (1)), and it is well known that the refractive index behaves in a peculiar manner in the neighborhood of an absorption band (anomalous dispersion). Thus, the use of the "normal" refractive index to compute the specular reflectivity is subject to some error. On the other hand, the accurate determination of specular reflectivity and the factors which determine it require single crystal measurements, but if such were available, one would usually prefer the transmission to the diffuse reflectance measurements.[37]

Thus, in conclusion, it does not appear that reflectance spectroscopy will ever "take over" transmission spectroscopy. It is best viewed as an alternative technique which is often capable of providing optical data where transmission spectroscopy fails, subject to the limitations mentioned above.

ACKNOWLEDGMENTS

I would like to express my appreciation to Dr. R. G. Giovanelli and Prof. Dr. G. Kortüm for supplying me with helpful correspondence and reprints of their numerous researches relating to reflectance spectroscopy.

REFERENCES

1. G. Kortüm and H. Schottler, Z. Elektrochem., **57**, 353 (1953).

2. G. Kortüm and J. Vogel, Z. Physik. Chem. (Frankfurt), **18**, 230 (1958).

3. E. Pitts, Proc. Phys. Soc., B, **67**, 105 (1954).

4. R. G. Giovanelli, Optica Acta, **2**, 153 (1955).

5. W. W. Wendlandt and H. G. Hecht, "Reflectance Spectroscopy," Interscience Publishers, New York, 1966.

6. A. Schuster, Astrophys. J., **21**, 1 (1905).

7. K. Schwarzschild, Göttinger Nachrichten, p. 41 (1906).

8. P. Kubelka and F. Munk, Z. Techn. Physik, **12**, 593 (1931).

9. D. B. Judd and G. Wyszecki, "Color in Business, Science, and Industry," 2nd ed., John Wiley & Sons, Inc., New York, 1963.

10. S. Chandrasekhar, "Radiative Transfer," Clarendon Press, Oxford, 1950 (Reprinted by Dover Publications, Inc., New York, 1960).

11. W. R. Blevin and W. J. Brown, J. Opt. Soc. Am. 52, 1250 (1962).

12. J. A. Van den Akker, L. R. Dearth, and W. M. Shillcox, J. Opt. Soc. Am., 56, 250 (1966).

13. D. G. Goebel, B. P. Caldwell, and H. K. Hammond, III, J. Opt. Soc. Am., 56, 783 (1966).

14. R. G. Giovanelli, Aust. J. Exptl. Biol. Med. Sci., 35, 143 (1957).

15. W. R. Blevin and W. J. Brown, J. Opt. Soc. Am., 51, 975 (1961).

16. G. Kortüm, W. Braun, and G. Herzog, Angew. Chem., Intern. Ed., 2, 333 (1963).

17. G. Mie, Ann. Physik, 25, 377 (1908).

18. G. V. Rozenberg, Bull. Acad. Sci. U.S.S.R., Phys. Ser. (English translation), 21, 1465 (1957).

19. W. R. Blevin and W. J. Brown, J. Opt. Soc. Am., 51, 129 (1961).

20. G. Kortüm and D. Oelkrug, Z. Naturforsch., 19a, 28 (1964); Die Naturwissenschaften, 23, 600 (1966).

21. E. A. Schatz, J. Opt. Soc. Am., 56, 389 (1966).

22. P. Kubelka, J. Opt. Soc. Am., 44, 330 (1954).

23. R. G. Giovanelli, "Diffusion Through non-uniform Media," in Progress in Optics, E. Wolf, Ed., Vol. II, North-Holland Publishing Company, Amsterdam, 1963, pp. 111-129.

24. R. G. Giovanelli, Austr. J. Phys., 10, 227 (1957).

25. R. G. Giovanelli, Austr. J. Phys., 12, 164 (1959).

26. P. R. Wilson, Austr. J. Phys., 13, 461 (1960); ibid., 14, 57 (1961); Mon. Not. R. Astr. Soc., 123, 287 (1962); 124, 383 (1962).

27. E. A. Schatz, J. Opt. Soc. Am., 57, 941 (1967).

28. G. Kortüm and G. Schreyer, Angew. Chem., 67, 694 (1955).

29. C. A. Lermond and L. B. Rogers, Anal. Chem., 27, 340 (1955).

30. R. W. Frei and M. M. Frodyma, Anal. chim. Acta, 32, 501 (1965).

31. W. J. Foote, Paper Trade J., 122, 35 (1946).

32. M. E. Everhard, D. A. Dickcius, and F. W. Goodhart, J. Pharm. Sci., 53, 173 (1964).

33. R. W. Frei, D. E. Ryan, and V. T. Lieu, Can. J. Chem., 44, 1945 (1966).

34. R. G. Giovanelli, Nature, 179, 621 (1957).

35. P. Rickard, F. Moss, and G. Roper, private communication.

36. P. D. Johnson, J. Opt. Soc. Am., 42, 978 (1952).

37. A. L. Companion, "Theory and Applications of Diffuse Reflectance Spectroscopy," in Developments in Applied Spectroscopy, Vol. 4, E. N. Davis, Ed., Plenum Press, New York, 1965.

38. N. T. Melamed, J. Appl. Physics, 34, 560 (1963).

THEORY OF SOME OF THE DISCREPANCIES OBSERVED IN APPLICATION OF THE KUBELKA-MUNK EQUATIONS TO PARTICULATE SYSTEMS

J. A. Van den Akker

Senior Research Associate and Chairman, Department of Physics and Mathematics, The Institute of Paper Chemistry, Appleton, Wisconsin 54911

INTRODUCTION

Beginning with the classic work of Stokes(1), a number of theories have been evolved during the past century to account for the observable optical properties of such light-scattering, light-absorbing media as piles of plates or films, translucent or partially opaque films, coatings, and sheets. Excellent reviews of the principal theories have been given by Judd and Wyszecki(2), Ingle(3), and Kubelka(4). The latter two have demonstrated the equivalence of some of the theories. Additionally, Kubelka(4) has presented a number of very useful equations stemming from the Kubelka-Munk theory(5).

The pioneering researches of Steele(6) and Judd(7), both of whom utilized the theory of Kubelka and Munk, resulted in special solutions and various mathematical aids that have been responsible, at least in part, for the widespread use of that theory. Today, many laboratories employ the Kubelka-Munk (K-M) theory for scientific or industrial purposes with generally good results. Computer programs based on the K-M theory are commonplace, and a number of laboratories having a long-term interest in the optics of turbid media have developed their own specialized graphical aids to the use of the theory. This popularity of the K-M theory is consonant with the essential soundness of the theory. Yet, over the years, significant discrepancies have been noted.

In our laboratories, fairly large departures from theoretical expectation have been noted in both the scattering and absorption coefficients of paper. In a very carefully conducted doctoral research, Foote[8] found that the scattering coefficient of dyed paper decreases with increasing absorption, the effect being small at low levels of dyeing, but pronounced at higher values of the absorption coefficient; he also found that, whereas the absorption coefficient increases linearly with the retained dye at low levels of dyeing, it falls away from the linear relationship at higher levels. In recent work conducted in Finland by Nordman, Aaltonen and Makkonen[9], Foote's finding on the dependence of the scattering coefficient on absorption was confirmed.

Probable reasons for the discrepancies are discussed in the present paper.

THEORY

Implications of the Theory

Adaptations of the K-M theory rest on the presumption that, in a homogeneous turbid medium, the scattering and absorption coefficients are constant throughout the system. The only spatial distribution of the radiation within the system that is compatible with this presumed constancy of the coefficients is that which is commonly designated as diffuse or isotropic; if the distribution were not diffuse, the coefficients would depend, among other things, on the form of the distribution function, and this would change from point to point throughout the system, always toward the diffuse form as the radiation penetrates more deeply into the system. Related to the implied diffuse radiation within the system is the condition that the system be illuminated diffusely. (A well-known problem is presented by systems having polished, flat surfaces at which reflection and refraction occur; even when external and internal reflection is allowed for, a nonideal condition regarding the spatial distribution of the radiation within the medium — near the boundary — exists.)

The first implication of the K-M theory, then, relates to the spatial distribution of the radiation within the medium, and it has been seen that the distribution must be diffuse. The second implication relates to the nature of the medium itself. The space of the ideal medium has a uniform absorption coefficient and is populated by an indefinitely large number of light-scattering centers, in each of which there is no absorption of radiation. The spatial distribution of these centers should be uniform, yet not sufficiently uniform (as in a crystal) to permit an observable diffraction pattern under unidirectional illumination. Perhaps the distribution can be compared to that of the molecules in Maxwell's model of an ideal gas: The number density and particle separation distribution function are uniform throughout the space. The mean separation of the light-scattering centers should be the smaller of the distances based on the following conditions: (a) The fractional absorption of radiation in traversing the mean distance $\underline{x}$ between particles, $1-\exp(-\underline{ax})$, can be expressed within acceptable limits of error as $-\underline{ax}$. This is equivalent to the condition of neglecting all terms in the series expansion of exp ($-\underline{ax}$) of degree higher than the first. (b) The mean separation of the light-scattering centers should be substantially smaller than the minimum linear dimension of the medium (e.g., the thickness in the case of a sheet or layer). Finally, for reasons given later, the absorption in a layer of the medium should be small as compared with the scattering power of that layer.

Discussion of Implications of the K-M Theory

Under circumstances where, of necessity, the medium must be illuminated unidirectionally, and the medium itself departs widely from the ideal described in the foregoing paragraph, special extension of the theory is clearly necessary. Undoubtedly the most notable of special developments is the theory of Duntley[10]. However, when this very useful theory is applied to certain media, like ordinary paper, for which the goniophotometric surface of the transmitted radiation is almost spherical (Lambertian) and possesses no observable lobe or spike, the application reduces logically to a mathematical model with added adjustable parameters.

In a recent attempt to arrive at an improved theory, Lathrop[11] considers the diffuse scattering of light from a planar array of particles on an absorbing "monolayer." Regarding the particles in a local region of diffuse radiation as each emitting radiation diffusely, he assumes that the flux per unit solid angle from each particle is independent of the polar angle θ (referred to the axis normal to the monolayer). He then integrates his expression for flux loss in the monolayer over a hemisphere to obtain the ratio of the K-M or Duntley absorption coefficient to the Bouguer coefficient. The ratio he obtains depends markedly on the absorption of the radiation in the monolayer, increasing as the absorption decreases, being about 2 when $\alpha\underline{W}=0.22$ ($\underline{W}$ being mass per unit area of the monolayer and α the Bouguer-Law absorption coefficient in reciprocal units of $\underline{W}$), about 6 at $\alpha\underline{W}=0.003$, and increasing indefinitely as $\alpha\underline{W}$ approaches zero. Lathrop's data for pigment films comprised of bonded colored glass particles (obtained by pulverizing filter glass in an alumina mortar and pestle and brushing through bronze-wire-sieve screens) display a dependence on $\alpha\underline{W}$ of the ratio of the absorption coefficients which is in fair agreement with his theory, when he selects an appropriate $\underline{W}$ for the monolayer. There is, however, a fundamental difficulty with the Lathrop theory and, as will be shown in the remaining sections of this paper, more plausible reasons for apparent failure of the K-M theory are to be found in departures of actual light-scattering systems from the ideal.

As pointed out above, Lathrop assumes that the flux per unit solid angle emitted by the particles in his planar array on a monolayer is independent of the polar angle. His $I(\theta)$, involved in the treatment leading to his Eqs. (6) and (7) and the theoretical lines in his Figs. 1-3, is uniform within a hemisphere. The central difficulty with the Lathrop theory is that $I(\theta)$ is not consistent with the diffuse radiation intensity function that appears in the differential equations of any valid theory relating intensity with position in the system. The diffuse intensity function should yield the flux emerging from dS, an elemental area in a plane that is normal to the reference direction of the system*, and originating in all the space beyond the elemental area as well as in the scattering centers in the plane. If the intensity of the diffuse radiation (defined in terms of an area in a plane normal to the reference axis) is i, the flux emerging from dS and contained in an elemental solid angle $d\omega$ at θ is

$$d\varphi = (i/\pi)\cos\theta\, dS\, d\omega. \tag{1}$$

Figure 1 presents the polar diagrams for the upward and downward moving radiation past a horizontal plane in the medium. If the radiation is diffuse, the diagrams are cross sections of spherical surfaces, a radius vector in either sphere being $d\varphi/dS\,d\omega = (i/\pi)\cos\theta$. Incidentally, Eq. (1) is quite consistent with the isotropic sensation of Lathrop's aviator in a cloud. It is only necessary to point out that the projection of the viewed dS on a plane normal to the line of sight is $dS\cos\theta$.

Equation (1) is employed in the next section in an analysis leading to the connection between the diffuse absorption coefficient and the Bouguer-Law coefficient.

Ratio of the Diffuse Absorption and Bouguer Coefficients

In this paper the independent variable of the K-M theory is W, the mass of a layer (or of the whole sheet) per unit area. The advantages of this choice over X (thickness) are appreciated where one is concerned with a compressible medium and its compositional aspects[14]. The coefficients of the theory are then the specific absorption and scattering coefficients, k and s. ($kW = KX$ and $sW = SX$, so that all charts and computer programs based on X, K and S can be readily employed.) In using W and the specific coefficients in theoretical discussions, it is of course, important to keep vertical spatial arrangement in mind.

*In this paper, the medium is visualized as a horizontal sheet or slab, illuminated diffusely about a vertical reference axis.

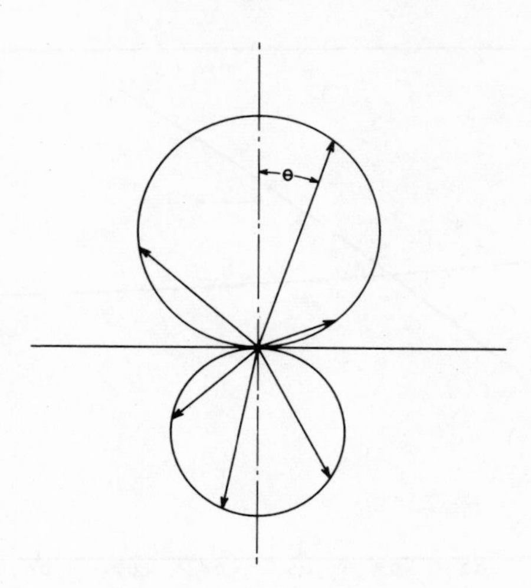

Figure 1. Spatial distribution of light in a light-scattering, light-absorbing medium (illuminated from below in illustration). A radius vector to either sphere is the energy per unit solid angle passing through unit area in a plane normal to the system axis.

We shall have occasion, in the remaining portions of this paper, to use effective and apparent coefficients. The effective absorption coefficient, to be used presently, is designated with a single prime and is defined by the equation,

$$A = 1-\exp(-k'W), \qquad (2)$$

for the absorption in a finite layer containing no scattering centers. The apparent coefficients, designated with a double prime, relate to the apparent discrepancies of the K-M theory which, in fact, are attributable to deviations of actual or assumed systems from the ideal. Their meaning will become clear in the context of the later analyses.

Calling Eqs. (1) and (2) into play, the absorption in a finite, clear layer that is diffusely illuminated is

$$1-\exp(-k'W) = (1/\pi) \int_0^{2\pi} \int_0^{\pi/2} \cos\theta \sin\theta [1-\exp(-\alpha W/\cos\theta)] d\theta d\varphi, \qquad (3)$$

in which α is the Bouguer-Law coefficient of absorption. This reduces to

$$\exp(-k'W) = 2(\alpha W)^2 \int_{\alpha W}^{\infty} [1/(z^3 \exp z)] dz. \qquad (4)$$

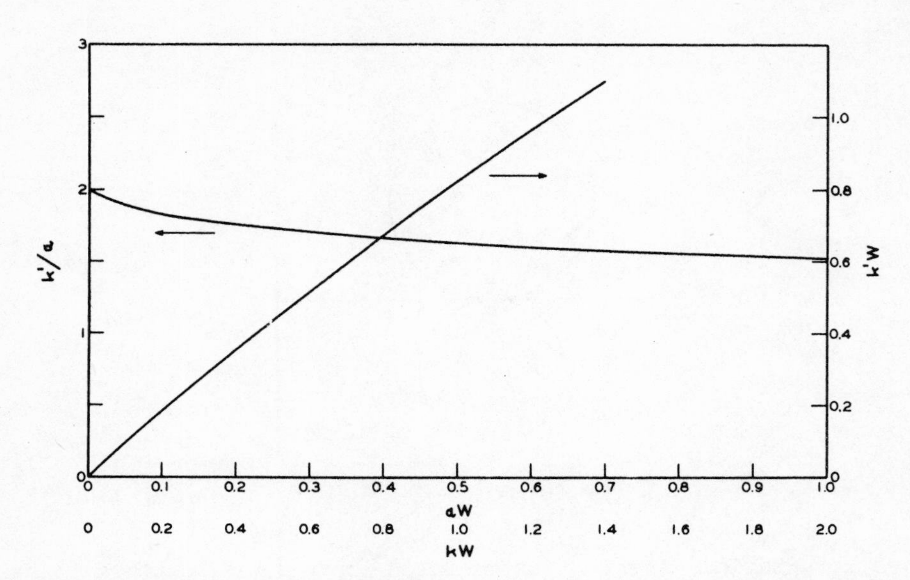

Figure 2. Theoretical dependence of the effective K-M absorption coefficient (relative to the Bouguer-Law coefficient) on the product of the Bouguer coefficient and the mass per unit area of a layer in an ideal medium.

The ratio k'/α calculated on the basis of Eq. (4) is presented in Fig. 2 for values of αW ranging from zero to unity. In his reference to the value 2.0 for this ratio, obtained by Kubelka[4] and Van den Akker[14], Lathrop[11] cited the earlier result out of context. The treatments referred to were conducted for the ideal medium, for which αW is the differential αdW, when, according to both the earlier and the present treatments, k'/α is indeed 2.0. When the absorption in a finite layer is considered, one finds the dependence of the ratio on αW to be that given in Fig. 2. This contrasts sharply with that given by Lathrop; as we have already pointed out, his function for the ratio yields indefinitely large values as αW approaches zero. Also presented in Fig. 2 is the dependence of $k'W$ on αW and kW, where k is the true K-M specific absorption coefficient (applicable to a differential layer in the ideal medium).

The decline of k' from 2α with increasing αW of a clear, absorbing layer is associated with the decreasing relative importance of the greater absorption along paths making larger angles with the axis of the system. When αW is very small, the flux transmitted through the layer in the angular range $d\theta$, at θ, is proportional to $\cos\theta\sin\theta$. The surface generated by rotating a polar plot of this function around the reference axis of the system displays a maximum or "lobe" at 45°. As the absorption increases, the lobe shifts to smaller angles, which means that the relative weight on transmission along shorter paths increases, thus accounting for the decrease of the effective absorption coefficient from 2α.

The shift in the spatial distribution of the radiation from the cosine law caused by appreciable absorption in a clear layer is to be noted. This, with its concomitant reduction of k' from the true k, constitutes one of the sources of apparent error of the K-M theory when applied to nonideal systems.

A very interesting consideration arises at this point. In discussing the ideal medium implied by the K-M theory, it was stated that "the absorption in a layer of the medium should be small as compared with the scattering power of that layer." In a medium of appreciable absorption coefficient, populated uniformly but thinly with light-scattering centers, the angular shift in the lobe discussed above could cause a reduction in k' from $k' = k$ at the diffusely illuminated surface of the system to k' significantly less than k at the other surface. It is obvious that this effect should become of decreasing importance as kW (of a layer or sheet) is made smaller in comparison with sW. In studies involving actual systems that would seem to simulate the ideal medium, this effect should be investigated.

Concentration Effects

Vertical Concentration of Scattering Centers. The particles of the ideal medium are distributed vertically to lie in a set of parallel, horizontal planes. The planes are arranged in closely-spaced pairs, the spacing being very small as compared with the spacing of the midplanes of the pairs. The medium is now sliced at the midplanes, thus forming a number of layers, each of which is a clear, absorbing layer of effective absorption coefficient k' with particles distributed uniformly over each face.

The mathematical slicing does not produce refracting and reflecting surfaces; but, of course, refraction and reflection of light could occur at the top and bottom surfaces of the system. In order that the treatment may be cleanly addressed only to the effects of vertical concentration of the particles, the index of refraction of the medium is equated to that of the surrounding space.

Each surface of the layers has reflectance r. Using an analysis very similar to that employed by Stokes[1], one finds for the reflectance of a single layer:

$$\underline{R}_o = \underline{r} + [\underline{r}(1-\underline{r})^2\exp(-2\underline{k}'\underline{W})]/[1-\underline{r}^2\exp(-2\underline{k}'\underline{W})]. \quad (5)$$

A similar analysis leads to the following equation for the transmittance of the layer.

$$\underline{T} = [(1-\underline{r})^2\exp(-\underline{k}'\underline{W})]/[1-\underline{r}^2\exp(-2\underline{k}'\underline{W})]. \quad (6)$$

These two equations are sufficient for the determination of the reflectivity $\underline{R}_\infty$ and, more important for our present purposes, the apparent absorption and scattering coefficients according to the K-M theory.

Charts and a number of useful equations may be called into play in connection with work of this sort[2,4].

Illustrative computations have been made on the basis of $\underline{r}$ = 0.1, which is of the right order for a number of real systems of interest. [The corresponding $\underline{sW}$ for the ideal medium (before concentrating the particles in planes) can readily be shown to be 0.2.] The results of the computations are given in Table I. Apparent values are designated with the double prime.

Table I. Apparent Coefficients and $\underline{R}_\infty$ for a Medium in Which the Particles Have Been Concentrated in the Surfaces of Layers

$\underline{k}'\underline{W}$	$\underline{sW}$	$\underline{k}''\underline{W}$	$\underline{s}''\underline{W}$	$\underline{R}_\infty$
0	0.2	0	0.223	1.0
0.01	0.2	0.010	0.223	0.742
0.05	0.2	0.050	0.223	0.518
0.5	0.2	0.483	0.244	0.173
1	0.2	0.93	0.295	0.122
2	0.2	1.80	0.45	0.102
3	0.2	2.60	0.66	0.102

Figure 3 presents the trends of the apparent absorption and scattering powers as functions of $\underline{k}'\underline{W}$.

Before discussing the trends, it should be pointed out that the scattering power at zero absorption is significantly greater than 0.2. This is a direct consequence of the concentration of the particles in two planes. The relative enhancement in $\underline{s}''\underline{W}$ diminishes with decreasing $\underline{r}$.

The reader concerned with colorants will be interested in noting that, for $\underline{sW}$ = 0.2, absorption is at the tinting level when $\underline{k}'\underline{W}$ = 0.01 (as judged by $\underline{R}_\infty$), at the level of moderate coloration when $\underline{k}'\underline{W}$ = 0.05, and at heavy levels when $\underline{k}'\underline{W}$ is 0.5 and higher.

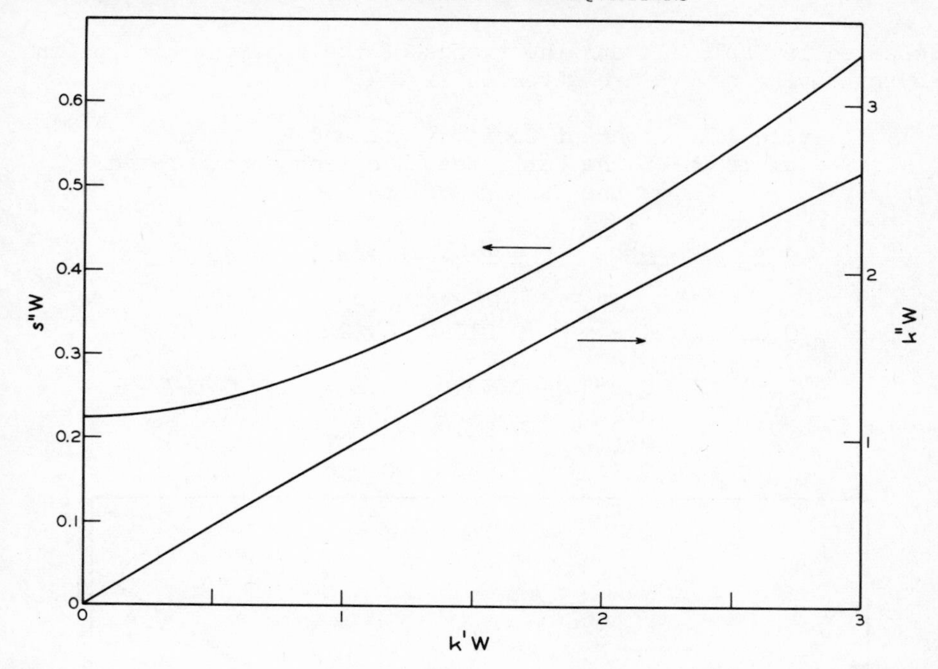

Figure 3. Theoretical dependence of the apparent K-M scattering and absorption powers on true effective absorption power when the light-scattering centers are concentrated vertically into the surfaces of layers formed by mathematical cuts.

The deviations of $s''W$ from constancy and $k''W$ from linearity with $k'W$ are insignificant until $k'W$ rises to $\overline{0.5}$ and higher. At very high values of $k'W$ there is a fall off in $k''W$ which is unrelated to the effect of absorption on the spatial distribution of the radiation (discussed in the preceding section); this decrement would be observed in a hypothetical system that could scatter light only along the system axis when under normal, unidirectional illumination.

Another type of layered system is obtained by making mathematical slices midway between the paired planes. Now one has a concentration of the particles at the middle of the layers. The expressions for R_0 and T are found to be

$$R_0 = 2[(r-r^2)\exp(-k'W)]/(1-r^2), \tag{7}$$

$$T = [(1-r)^2\exp(-k'W)]/(1-r^2). \tag{8}$$

It is to be remembered that the particles at the middle of the layer lie in a pair of closely-spaced parallel planes.

The results of the computations (again based on $r = 0.1$) are presented in Table II, and the trends of the apparent absorption and scattering powers are given in Fig. 4.

Table II. Apparent Coefficients and R_∞ for a Medium in Which the Particles Have Been Concentrated at the Centers of Layers

$k'W$	sW	$k''W$	$s''W$	R_∞
0	0.2	0	0.223	1.0
0.01	0.2	0.010	0.223	0.742
0.05	0.2	0.050	0.221	0.517
0.50	0.2	0.509	0.205	0.147
1	0.2	1.03	0.178	0.074
2	0.2	2.07	0.107	0.025
3	0.2	3.07	0.0565	0.0091

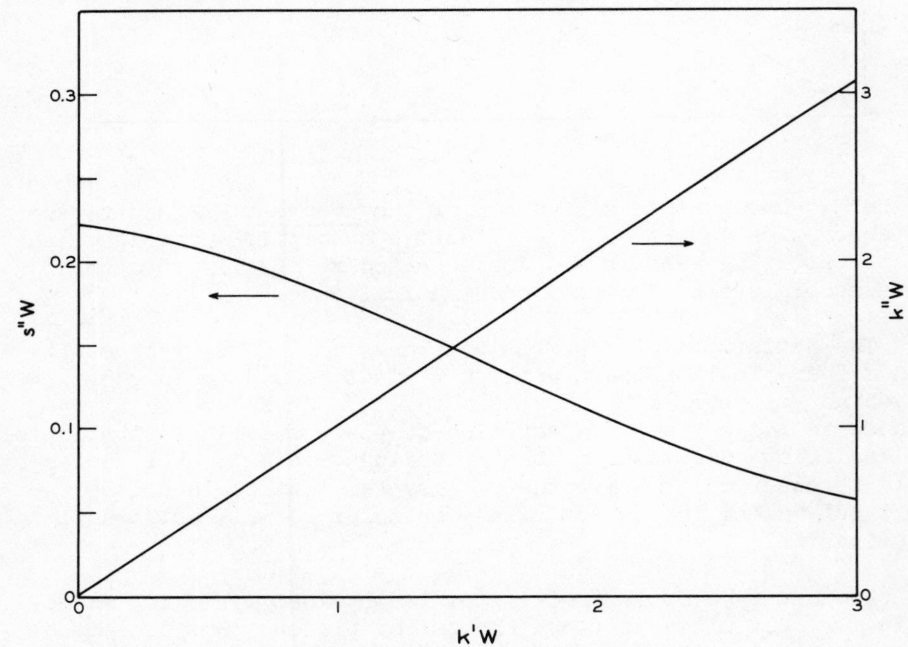

Figure 4. Theoretical dependence of the apparent K-M scattering and absorption powers on true effective absorption power when the light-scattering centers are concentrated vertically into mid-planes of layers formed by mathematical cuts.

Again, the deviations of $s''W$ from constancy and $k''W$ from linearity with $k'W$ are not significant until $k'W$ rises to 0.5 and higher. Now, however, the apparent scattering power *decreases* with increasing absorption.

Although this hypothetical system would not seem to simulate actual media of practical interest, its behavior, along with that of the first system, shows that vertical concentration of light-scattering centers can modify the apparent K-M coefficients – and that the direction of the changes depends on the manner of concentration of the centers.

In both of these systems, the dependence of the apparent coefficients on absorption is weak for reflectivities in the ordinary practical range but strong for reflectivities substantially less than about 0.15.

Unrelated Horizontal Concentrations of Scattering Centers and Absorption Areas. The first of the hypothetical systems considered above is used as a starting point for several horizontal distributions. In the first (the remaining ones are treated in later sections), we imagine that the scattering centers are distributed only in a number of random patches whose total area is a fraction η_1 of the whole area. The reflectance of either surface of these patches is now r/η_1. The absorbing material is distributed only in a number of unrelated random patches whose total area is a fraction η_2 of the whole area. The absorption in each patch depends on $k'W/\eta_2$, which one can regard as the result of concentrating absorbing molecules in an absorption-free substance, in which case k' is increased, or of simply concentrating the absorbing material (with increase of thickness), so that k' remains constant and the mass per unit area increases locally. In any case, the mass per unit area of the whole layer is regarded as remaining constant. The fraction of the whole area occupied by scattering centers only is $\eta_1(1-\eta_2)$; that which is occupied by absorbing area only is $\eta_2(1-\eta_1)$; that which is occupied by both scattering and absorbing areas is $\eta_1\eta_2$; and that which is occupied by neither is $(1-\eta_1)(1-\eta_2)$. Analysis of this system leads to the following equations for R_o and T.

$$R_o = (1-\eta_2)\left[r + \frac{r(1-r/\eta_1)^2}{1-(r/\eta_1)^2}\right] + \eta_2\left[r + \frac{r(1-r/\eta_1)^2\exp(-2k'W/\eta_2)}{1-(r/\eta_1)^2\exp(-2k'W/\eta_2)}\right], \quad (9)$$

$$T = \eta_1(1-\eta_2)\frac{(1-r/\eta_1)^2}{1-(r/\eta_1)^2} + \eta_2(1-\eta_1)\exp(-k'W/\eta_2) + \eta_1\eta_2\frac{(1-r/\eta_1)^2\exp(-k'W/\eta_2)}{1-(r/\eta_1)^2\exp(-2k'W/\eta_2)} + (1-\eta_1)(1-\eta_2). \quad (10)$$

In the interest of comparison, $\underline{r}$ has again been put equal to 0.1, and illustrative computations have been based on $\eta_1 = \eta_2 = 0.5$. The results are given in Table III.

Table III. Apparent Coefficients and $\underline{R}_\infty$ for Unrelated Horizontal Concentrations of Scattering and Absorbing Areas

$\underline{k}'\underline{W}$	$\underline{s}\underline{W}$	$\underline{k}''\underline{W}$	$\underline{s}''\underline{W}$	$\underline{R}_\infty$
0	0.2	0	0.200	1.0
0.001	0.2	0.0077	0.203	0.76
0.01	0.2	0.0172	0.203	0.67
0.05	0.2	0.0538	0.202	0.49
0.25	0.2	0.214	0.214	0.27
0.5	0.2	0.351	0.230	0.207
1	0.2	0.511	0.260	0.173
2	0.2	0.600	0.280	0.163
3	0.2	0.613	0.283	0.162

It will be seen that, by chance, the choice of $\eta_1 = 0.5$ results in an absorption-free apparent scattering power close to the ideal. The unrelated positioning of scattering and absorption areas leads to a curious trend in the apparent absorption coefficient at very low levels of absorption — where the diffusion of the radiation is ideal and where the absorption can be written, with good accuracy, $1\text{-}\exp(-\underline{k}'\underline{W}) = \underline{k}'\underline{W} = \underline{k}\underline{W} = 2\alpha$. The ratio $\underline{k}''\underline{W}/\underline{k}'\underline{W}$ is 7.7 at $\underline{k}'\underline{W} = 0.001$, 1.72 at 0.01, and diminishes to substantially less than unity at higher levels of absorption. The trend in $\underline{s}''\underline{W}$ is weak at moderate levels of absorption.

When the scattering and absorption occur in the <u>same</u> concentrated regions, the trends are interestingly modified. This is covered in the following paragraphs.

<u>Related Horizontal Concentrations of Scattering and Absorbing Areas</u>. In many filamentous and particulate systems the scattering and absorption occur in the same regions. It is, therefore, of considerable interest to study the effects of concentrating the scattering and absorbing areas into the same patches, the total area of which constitutes a fraction η of the whole area. The patches are considered to be distributed randomly. Analysis of this system leads to Eqs. (11) and (12) for $\underline{R}_o$ and $\underline{T}$.

$$\underline{R}_o = \underline{r} + \frac{\underline{r}(1-\underline{r}/\eta)^2\exp(-2\underline{k}'\underline{W}/\eta)}{1-(\underline{r}/\eta)^2\exp(-2\underline{k}'\underline{W}/\eta)}, \tag{11}$$

$$\underline{T} = (1-\eta) + \eta \frac{(1-\underline{r}/\eta)^2\exp(-\underline{k}'\underline{W}/\eta)}{1-(r/\eta)^2\exp(-2\underline{k}'\underline{W}/\eta)}. \tag{12}$$

Illustrative calculations, based on $\underline{r} = 0.1$ and $\eta = 0.5$, are presented in Table IV and Fig. 5.

Table IV. Apparent Coefficients and $\underline{R}_\infty$ for Related Horizontal Concentrations of Scattering and Absorbing Areas

$\underline{k}'\underline{W}$	sW	$\underline{k}''\underline{W}$	$\underline{s}''\underline{W}$	$\underline{R}_\infty$
0	0.2	0	0.200	1.0
0.01	0.2	0.010	0.200	0.73
0.05	0.2	0.048	0.192	0.500
0.10	0.2	0.091	0.186	0.386
0.25	0.2	0.200	0.176	0.248
0.50	0.2	0.323	0.170	0.178
1	0.2	0.450	0.176	0.145
2	0.2	0.518	0.185	0.134
3	0.2	0.522	0.186	0.134

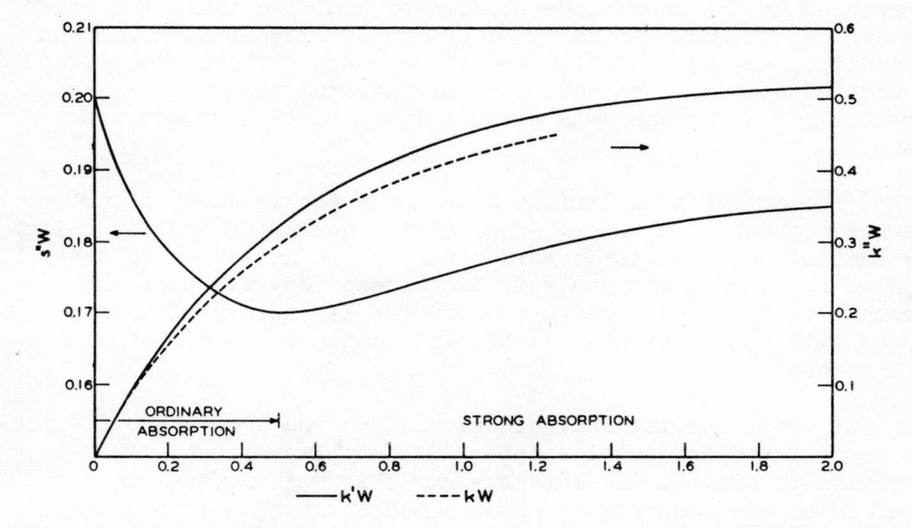

Figure 5. Theoretical dependence of the apparent K-M scattering and absorption powers on true effective absorption power when the scattering centers and absorbing material are concentrated into the same, randomly distributed patches (see text). The case illustrated is that for which the total area of the random patches is one-half the whole area.

In the moderate range of absorption (at which so much experimental work is done), this system yields significant changes in the apparent $\underline{s}''\underline{W}$ from constancy and $\underline{k}''\underline{W}$ from linearity with $\underline{k}'\underline{W}$. More important, the direction and magnitude of the trends are in approximate agreement with the experimental findings referred to in the introductory comments. From $\underline{k}'\underline{W} = 0$ to the fairly strong level of 0.5 (at which a thin sheet of ten layers of the medium would have a transmittance of only 0.01), $\underline{s}''\underline{W}$ falls off by about the same percentage as that observed by Foote[8] and Nordman[9] on dyeing paper to substantial levels of absorption. While a quantitative comparison of the departure from linearity of $\underline{k}''\underline{W}$ with $\underline{kW}$ cannot be made with that found experimentally by Foote, comparison of the dashed curve in Fig. 5 with the data of Foote shows that the shapes are very similar. The dashed curve, which includes the departure of the effective $\underline{k}'\underline{W}$ from $\underline{kW}$ (and 2α) — discussed early in this paper — is of interest where one has information on the Bouguer-Law coefficient. When the data of Lathrop[11] are translated to the kind of plot given in Fig. 5, curves of similar shape are obtained. Duysens[12] has considered the mutual screening of pigment molecules in particles present in a suspension. His concern was the flattening of absorption spectra on going from pigments in solution to particles in suspension. Perhaps the hypothetical system considered in this section would more nearly correspond to the phenomenon studied by him than would the other systems considered. Actually, there are two types of screening. In addition to that to which reference has just been made, there is the reduction in scattering resulting from absorption in a particle[13]. The analysis given in this section covers both cases.

Agreement between limited data and a theory does not prove the correctness of a model. Challenging questions must be raised, and crucial experiments performed before one may conclude that the type of system and the analysis just presented will properly account for observed departures from the K-M theory of data observed with actual porous, particulate media which are simulated by the model.

The model and its analysis should provide a more valid explanation of the discrepancies observed in application of the K-M theory to porous, particulate systems than mathematical models based on ad hoc assumptions about the distribution of radiation within a medium. An experimental program based on the theory of this section is being planned.

Intensive Concentration of the Absorbing Matter Only. If one imagines that the absorbing centers of the layered system are concentrated into random patches of total area occupying a fraction η of the whole area, with the scattering centers uniformly distributed over the faces of the layers and the associated mass per unit area $\underline{W}$ everywhere the same, the absorption in a differential layer in a patch would be $(\underline{k}/\eta)\underline{dW}$. The superficial view is that the average absorption in a layer is unaffected by the concentration; this is based on the consideration that the fraction of radiation falling on the absorbing patches is η, and that $\eta(\underline{k}/\eta)\underline{dW}$

= $\underline{k}d\underline{W}$. This is probably a good approximation for mild concentration of a weakly absorbing substance occurring in finite layers. However, as the degree of concentration intensifies, the approximation, $1-\exp(-\underline{k}'\underline{W}/\eta) \cong -\underline{k}'\underline{W}/\eta$, breaks down badly, even for mildly absorbing material.

In certain powdery mixtures of predominantly white particles with a low percentage of very small particles of a soluble, highly absorbing substance, the reflectivity is very much higher than that computed on the basis of the K-M theory, on taking into account (improperly) the absorbing power of the dark particles. When the mixture is treated with a solvent, the absorbing matter dissolves and diffuses throughout the mixture, resulting in low reflectivity[14]. Certain natural materials, like wood, may be light-colored, yet contain highly absorbing particles. Darkening occurs on subjecting these materials to certain solvents. Recently, Gupta and Mutton[15], have concluded a very interesting study of the color of wood and groundwood, in which the importance of the effect was demonstrated.

In isotropic diffusion, the spheres of influence of the dissolved absorbing substance grow with time, and overlap. If the regions are cut by parallel planes (to form our layered model), one visualizes growing patches of absorbing matter. The concentration of absorbing matter into patches that was described in the penultimate paragraph above is not the reverse of a diffusion process, primarily because of the overlapping of patches and variability of the absorption coefficient in the latter. Nevertheless, the principal ideas and effects can be brought out by a treatment of the envisioned system, in which the absorbing centers are systematically crowded into smaller and smaller areas.

Analysis similar to that employed in earlier sections yields Eqs. (13) and (14) for $\underline{R}_o$ and $\underline{T}$. Again, $\underline{r} = 0.1$. Table V presents

$$\underline{R}_o = (1-\eta)\left[\frac{2(\underline{r}-\underline{r}^2)}{1-\underline{r}^2}\right] + \eta\left[\underline{r} + \frac{\underline{r}(1-\underline{r})^2\exp(-2\underline{k}'\underline{W}/\eta)}{1-\underline{r}^2\exp(-2\underline{k}'\underline{W}/\eta)}\right], \qquad (13)$$

$$\underline{T} = (1-\eta)\,\frac{(1-\underline{r})^2}{1-\underline{r}^2} + \eta\left[\frac{(1-\underline{r})^2\exp(-\underline{k}'\underline{W}/\eta)}{1-\underline{r}^2\exp(-2\underline{k}'\underline{W}/\eta)}\right]. \qquad (14)$$

the computed values of the apparent K-M absorbing power $\underline{k}''\underline{W}$, scattering power $\underline{s}''\underline{W}$, and $\underline{R}_\infty$ for eight widely different degrees of concentration at three levels of $\underline{k}'\underline{W}$. The dependence of $\underline{R}_\infty$ on η is presented in Fig. 6.

Table V. Apparent Coefficients and $\underline{R}_\infty$ at Various Levels of Concentration of Absorbing Matter Only

η	$\underline{k}'\underline{W}$	$\underline{s}''\underline{W}$	$\underline{k}''\underline{W}$	$\underline{R}_\infty$
1.0	0.01	0.223	0.0100	0.742
0.5	0.01	0.223	0.0099	0.743
0.2	0.01	0.223	0.0099	0.743
0.05	0.01	0.223	0.00908	0.753
0.01	0.01	0.223	0.00595	0.795
0.001	0.01	0.223	0.000904	0.914
0.0001	0.01	0.223	0.000090	0.972
0.00001	0.01	0.223	0.0000090	0.991
1.0	0.05	0.223	0.050	0.518
0.5	0.05	0.223	0.048	0.524
0.2	0.05	0.223	0.045	0.536
0.05	0.05	0.223	0.0304	0.597
0.01	0.05	0.223	0.00903	0.754
0.001	0.05	0.223	0.000904	0.914
0.0001	0.05	0.223	0.000090	0.972
0.00001	0.05	0.223	0.0000090	0.991
1.0	0.50	0.244	0.483	0.173
0.5	0.50	0.254	0.369	0.214
0.2	0.50	0.244	0.185	0.312
0.05	0.50	0.228	0.0370	0.533
0.01	0.50	0.223	0.00908	0.753
0.001	0.50	0.223	0.000904	0.914
0.0001	0.50	0.223	0.000090	0.972
0.00001	0.50	0.223	0.0000090	0.991

The most dramatic effect of sequestration of the absorbing matter is seen at the highest level of $\underline{k}'\underline{W}$ where, on increasing the concentration 1000-fold, the apparent absorption coefficient (Table V) and reflectivity (Table V and Fig. 6) approach levels corresponding to a high level of luminous reflectivity; a further 100-fold enhancement (to $\eta = 0.00001$) yields almost the maximum reflectivity attainable in white materials.

When the absorption is sufficiently low, an appreciable but not extreme concentration of the absorbing matter produces little effect. For example, when $\underline{k}'\underline{W} = 0.01$, $\underline{k}''\underline{W}$ and $\underline{R}_\infty$ are not significantly modified by a five-fold increase in concentration; in this range the approximation $1\text{-}\exp(-\underline{k}'\underline{W}/\eta) \cong \underline{k}'\underline{W}/\eta$ holds with good accuracy. However, the concentrated patches, even at the moderate level of $\underline{k}'\underline{W} = 0.01$, become opaque when the concentration enhancement reaches 1000-fold ($\eta = 0.001$), and $\underline{k}''\underline{W}$ and $\underline{R}_\infty$ become the same as the corresponding values for all

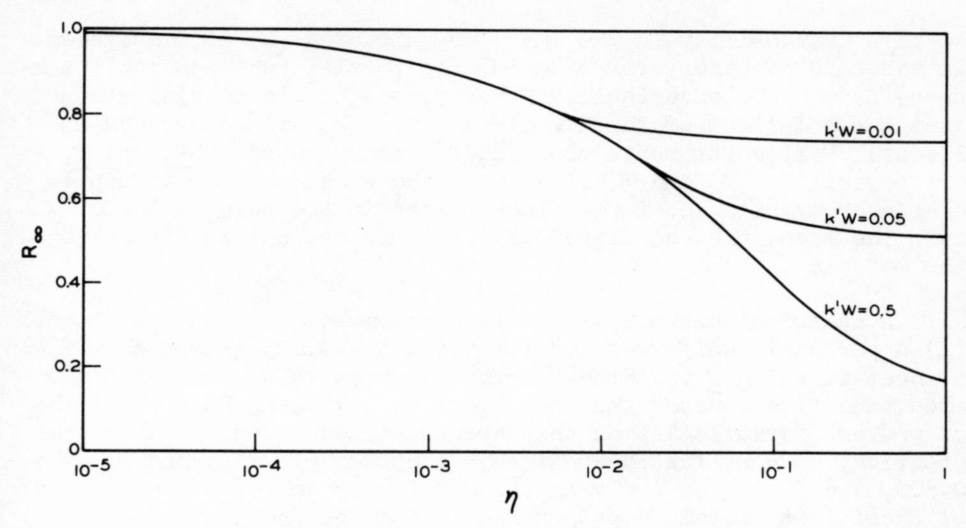

Figure 6. The theoretical effect on reflectivity, for three levels of absorption, of concentrating the absorbing matter alone into smaller and smaller randomly distributed patches. At the lowest level of absorption, a ten-fold concentration scarcely affects $\underline{R}_\infty$, whereas the same concentration enhancement results in a large change where the original absorption power was fairly high. When the concentration proceeds beyond 100-fold, the patches at all levels of absorption attain the same limiting absorbance, and the three curves fuse together. When the sequestration is carried far enough, a very high level of reflectivity is attained.

higher levels of absorption. The approximation just referred to then breaks down badly: Absorption in a patch is, of course, limited to unity $[1-\exp(-\underline{k}'\underline{W}/\eta) \rightarrow 1.0]$, whereas the approximation form increases linearly with $1/\eta$ to 10 when $\eta = 0.001$, and to 1000 when $\eta = 0.00001$.

For reasons given earlier, the hypothetical changes discussed above do not correspond accurately to the reverse of solution and diffusion of darkly-colored particles distributed in a lightly-colored system. However, the treatment improves one's understanding of the optical phenomena that occur when diffusion of absorbing matter takes place, and the theory, therefore, is heuristic.

Nonuniform Thickness

A significant source of error in the determination of K-M scattering and absorption coefficients is attributable to point-to-point variability in the mass per unit area (W) of the system. In the case of paper, for example, the local W (whole sheet thickness) can be quite variable. Error arises in the consideration that the solutions of the K-M differential equations are nonlinear. Most instruments employed for measurement of R_∞ and R_o or, better[16], R_∞ and T, illuminate the specimen over an appreciable area; because of the fluctuations in the point value of W over the area, the calculated coefficients are not intrinsic to the material.

A numerical example will suffice to make the point. Consider (A) a perfectly uniform sheet for which $W = 0.005$ g/cm^2, and (B) a sheet in which W is three-fourths the mean value over half the area, and five-fourths the mean over the remaining half (sheets of poorer "formation" than this are sometimes encountered in practice). When, for Sheet A, $sW = 1.25$, $kW = 0.312$, and $R_\infty = 0.500$, one finds, on applying the K-M theory to the two half areas of Sheet B (to obtain a proper mean T for the whole area), that the apparent sW and kW are, respectively, 1.20 and 0.300, i.e., 4% less than the proper intrinsic values. The nature of this source of discrepancy is such that the error increases very rapidly with increasing degrees of "wildness" of formation, but diminishes to insignificance as the formation improves to superior levels (at which, however, the nonuniformity is still easily measured).

Light-Trapping

A long-recognized phenomenon in certain types of media is that of light-trapping, a consequence of total internal reflection in transparent bodies having optically smooth surfaces. Light-trapping provides another deviation of actual systems from the ideal medium implied by the K-M and other theories. It is believed to be of possible importance in paper and other filamentous sheets, particularly where, as in paper, there is optical contact at points of bonding or intimate contact between the fibers (or filaments). Because of the undulations in the surfaces of natural fibers, some degree of light-trapping may be expected in a paper sheet; additionally, the role played by the "windows" between fibers provided by the bonds should be of importance in view of the substantial level of "relative bonded area" typically found in paper. In a Special Studies program, Mr. Philip F. Brown, graduate student at The Institute of Paper Chemistry, has designed and constructed a solid integrating cavity which should be useful in research on the effects of light-trapping in paper and other filamentous materials[17].

CONCLUSIONS

Starting with the diffusely illuminated ideal medium implied by the Kubelka-Munk theory, several hypothetical light-scattering, light-absorbing systems have been set up by (a) concentrating the light-scattering centers into the surfaces of layers (defined by mathematical cuts), (b) concentrating the light-scattering centers into the midplanes of layers, (c) horizontally concentrating the light-scattering centers and absorbing matter of the system (a) above into unrelated random patches, (d) horizontally concentrating the light-scattering centers and absorbing matter into the same randomly distributed patches, and (e) horizontally concentrating only the absorbing matter of system (a) above into patches of diminishing total area. Analysis of these systems shows that, in all cases, deviations from the ideal medium produce changes in the dependence of the apparent Kubelka-Munk scattering and absorption coefficients on the true absorption coefficient; in the case of the apparent scattering coefficient, the change can be monotonically up or down, or nonmonotonic, while the apparent absorption coefficient falls away from linearity with the true coefficient in all cases but one. It is observed that system (d) above yields changes in the apparent Kubelka-Munk coefficients that are most nearly in accord with laboratory data. Furthermore, this system most closely simulates the type of porous and filamentous or particulate systems on which the observations leading to this study have been made. Accordingly, the principal conclusion of this work is that the departures from the Kubelka-Munk theory can most properly be accounted for through consideration of the departures of actual media from the ideal system, and that the theory presented for system (d) above is expected to be most in accord with observations when the medium is porous and filamentous or particulate.

Also of interest are several other reasons for departures of observations from the Kubelka-Munk theory, namely, optical sequestration, nonuniformity of thickness of the medium, and light-trapping. Of these three effects, the first two might be of importance in all types of media, while it is expected that light-trapping would be most important in filamentous media.

LITERATURE CITED

1. G. G. Stokes, Proc. Roy. Soc., London, 11:545 (1860-1862).

2. Deane B. Judd, and Günter Wyszecki, "Color in Business, Science and Industry," 2d ed., Wiley, New York, (1963), 500 p.

3. G. W. Ingle, ASTM Bull. 116:32 (1942).

4. P. Kubelka, J. Opt. Soc. Am. 38:448 (1948).

5. P. Kubelka, and F. Munk, Z. tech. Physik. 12:593 (1931).

6. F. A. Steele, Paper Trade J. 100 (12):37 (March 21, 1935).

7. D. B. Judd, J. Res. Natl. Bur. Stds. 19:287 (1937).

8. W. J. Foote, Tech. Assoc. Papers (TAPPI), Series XXII:397 (1939).

9. L. Nordman, P. Aaltonen, and T. Makkonen, In "Transactions of the Symposium on Consolidation of the Paper Web," p. 909, Tech. Sect. Brit. Paper and Board Makers' Assoc., London, (1966).

10. S. Q. Duntley, J. Opt. Soc. Am. 32:61 (1942).

11. A. L. Lathrop, J. Opt. Soc. Am. 56:926 (1966).

12. L. N. M. Duysens, Biochim. et Biophys. Acta 19:1 (1956).

13. J. A. Van den Akker, In "Transactions of the Symposium on Consolidation of the Paper Web," p. 948, Tech. Sect. Brit. Paper and Board Makers' Assoc., London, (1966).

14. J. A. Van den Akker, Tappi 32:498 (1949).

15. V. N. Gupta, and D. B. Mutton, Pulp Paper Mag. Can. 68:T107 (1967).

16. J. A. Van den Akker, "Theory of the Optical Properties of Pulp," TAPPI Monograph Series 27:17 (1963).

17. P. F. Brown, Unpublished work, 1966-67.

DIFFUSE REFLECTANCE SPECTROSCOPY - A QUANTITATIVE TECHNIQUE FOR CHARACTERIZING LIGAND FIELD SPECTRA

Edward A. Boudreaux and Joseph P. Englert

Louisiana State University in New Orleans

New Orleans, Louisiana 70124

The theory of diffuse reflectance spectroscopy of compacted powders was placed on a quantitative basis by Melamed (1), who considered the total scattering transmission and absorption of radiation from individual spherical particles. In spite of the limitations prevalent in a spherical particle model, Melamed was able to show that the absorption coefficient could be evaluated quantitatively from diffuse reflectance, but the errors could be as great as 25% even when the molar extinction coefficient was large.

This work reports a limited study to determine the quantitative reliability in obtaining absorption coefficients from diffuse reflectance spectra for coordination complexes exhibiting ligand field bands of low molar extinction coefficients. The study requires that the test samples be of such a structural constitution that they retain their molecular identity in solution and in the solid. For these purposes $K_3Fe(CN)_6$ and $NiSO_4 \cdot 7H_2O$ were selected which contain the chromophores $Fe(CN)_6^{3-}$ and $Ni(OH_2)_6^{2+}$ in the solid and in aqueous solution respectively.

EXPERIMENTAL

Analytical Reagent grade samples of $K_3Fe(CN)_6$ and $NiSO_4 \cdot 7H_2O$ were employed without further purification. The samples were ground to uniform particle size in a high speed pulveriser with agate grinding balls.

Grinding was conducted over an 8-12 hour period at intermittent intervals of one hour, since the pulveriser could only be set for a maximum grinding time of one hour. The powder was subjected to visual microscopic examination at the end of each third or fourth grinding interval, so as to determine when particle size uniformity was attained. An average particle size of 12μ was obtained for $K_3Fe(CN)_6$ within an 8 hour interval while that of $NiSO_4 \cdot 7H_2O$ was 28μ after 12 hours of grinding. In both cases the maximum estimated error was ±10-15%.

Diffuse reflectance spectra were recorded with a Beckman DK-1 spectrophotometer equipped with a reflectance attachment using fresh MgO as a reference. The $NiSO_4 \cdot 7H_2O$ sample was run in pure form but the $K_3Fe(CN)_6$ had to be diluted 1:1 by volume in MgO because of the greater intensity of the absorption bands. In all recordings of the spectra numerous scans were made to insure the reproduciability and accuracy of the data.

The required values of refractive indices, n, were obtained from standard handbooks, except for the $K_3Fe(CN)_6$-MgO mixture which was estimated by weighting the refractive indices of the pure components to their respective molar volumes, thus giving an average value $\bar{n}$.

Absorption spectra of the aqueous solutions were also obtained on the DK-1 using $10cm^3$ quartz sample cells of 1 cm path length. These cells were mounted forward of the integrating sphere. Solutions having concentrations of $5X10^{-4}$ for $K_3Fe(CN)_6$ and 0.2650M for $NiSO_4 \cdot 7H_2O$ were employed.

THEORY AND COMPUTATIONAL PROCEDURE

The Melamed equation is a function of average reflection coefficients for internally and externally incident radiation, m_i and m_e respectively; which vary with refractive index, and has the form

$$R = \frac{2Xm_e + X(1-2Xm_e)\,(1-m_eR)\,T}{(1-m_eR) - (1-X)\,(1-m_e)\,TR}$$

where T, the fraction of transmitted radiation, is given by $(1-m_i)M/1-m_iM$. The scattering factor, X, has the value 0.284/1-0432T for close packed spherical particles, and

$$M = \frac{2}{(kd)^2} \left[1 - (kd+1)e^{-kd}\right]$$

which may be expanded for low values of kd to yield M= 1-2/3 kd. In this equation k is the optical absorption coefficient and d the average particle diameter. Finally R is the absolute diffuse reflectance which is derivable from the observed relative reflectance, Ro, by correcting for the absolute reflectance of the MgO reference. The reflection coefficients, m_i and m_e were obtained from the data provided by Melamed (1).

The solution of this equation was programmed for the IBM-1620 computer. Required input data are the values of x, m_e, m_i, d, and the observed values of Ro. The program initially executes a subroutine which converts Ro to R. It then computes values of T and k and outputs the k values. There is in addition a subroutine which produces a rough plot of k vs. λ. This routine proceeds in the following manner.

a. Test each k value and adjust to a maximum relative scale.
b. The number of intervals within the maxima and minima must be specified for λ.
c. The starting (lowest) λ value is inputed.
d. The number of points desired within each λ interval must be specified.

The overall results for the computed curve can be substantially improved if displayed on a graphical plotter thus providing a smooth and continous curve.

RESULTS AND CONCLUSIONS

Both reflectance and solution spectral curves are given in Figures 1 and 2. More detailed data are provided in Tables I and II.

Qualitatively speaking the reflectance curves duplicate those of absorption from solution quite well. However, the relative band intensities and band shapes are somewhat modified on going from the absorption of the solution to the diffuse reflectance of the solid.

TABLE I

SPECTRAL DATA FOR $K_3Fe(CN)_6$

SOLUTION ($5x10^{-4}M$)

(mμ)	%T	A		k
258	27	0.569	1130	1.3
300	18	0.745	1490	1.7
315(sh)	28	0.553	1100	1.3
415	31	0.509	1000	1.2

SOLID (1:1 in MgO)

(mμ)	R	$(kd)x10^{-6}$	$\bar{n}=1.58$	k
254	0.312	0.37	d=28μ	1.3
305	0.260	0.50		1.8
315(sh)	0.255	0.51		1.8
417	0.180	0.85		3.0

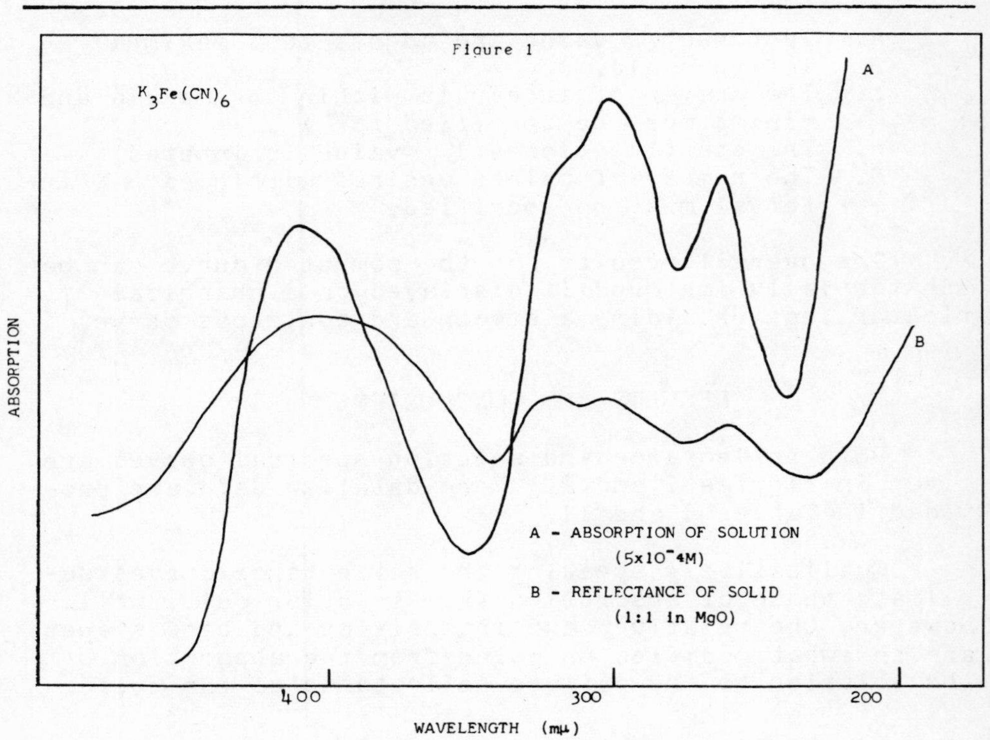

TABLE II
SPECTRAL DATA FOR $NiSO_4 \cdot 7H_2O$

SOLUTION (0.265M)

(mμ)	%T	A		k
395	3	1.523	4.6	2.8
650	32	0.495	1.9	1.2
710	27	0.570	2.2	1.3
1176	26	0.588	2.1	1.3

SOLID (pure)

(mμ)	R	(kd)x10^{-6}	n = 1.49	k
380	0.525	0.25	d = 12μ	2.1
650	0.595	0.15		1.3
680	0.590	0.16		1.3
1170	0.540	0.18		1.5

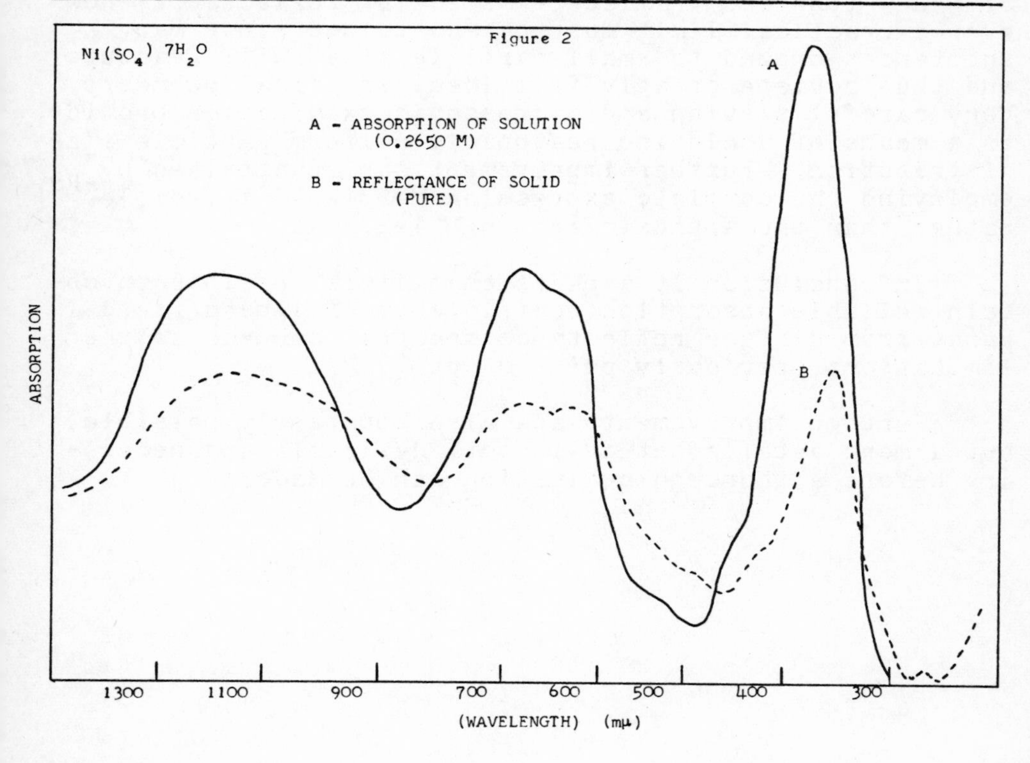

The wavelengths of the absorption maxima are also slightly shifted in going from one medium to the other.

A close inspection of Tables I and II suggests that the k values obtained from solution and from the solid should be in better agreement for $K_3Fe(CN)_6$ than for $NiSO_4 \cdot 7H_2O$, since the former has molar extinction coefficients nearly 10^3 greater than those of the latter. However this is apparently not the case, and in fact there appears to be generally better duplication of k values for $NiSO_4 \cdot 7H_2O$ than for $K_3Fe(CN)_6$. Hence it cannot be concluded that because the molar extinction coefficient is greater that the evaluation of absorption coefficients should necessarily be more accurate.

More specifically the results show that for diffuse reflectance in the range $0.2<R<0.7$, and kd in the interval $0.1<kd<0.5$, the maximum error in accurately reproducing k is about 1 to 25%. However as kd approaches unity R appears to be relatively insensitive to variations in k, and apparently for $kd>1$ this situation would become progressively worse.

The most critical factors to control are actual particle shape and size. Small particles in the range of 5-20μ tend to minimize the specular reflectance, however a practical limit must be maintained since many substances ground to small particle sizes will fracture and thus deviate greatly from ideal spherical geometry. Very careful sieving and microscopic examination provides a means of achieving reasonably uniform particle size distribution. Further improvement can be attained by employing the complete expression $M=2/(kd)^2[1-(kd+1)e^{-kd}]$ rather than the approximate one; $M=1-2/3\ kd$.

In conclusion it appears that it is possible to obtain reliable absorption coefficients of ligand field bands from diffuse reflectance spectra, subject to the limitations previously pointed out.

Further improvements are also apparently possible, but a more detailed study is both desirable and necessary before a thorough evaluation can be made.

HIGH TEMPERATURE REFLECTANCE SPECTROSCOPY AND DYNAMIC REFLECTANCE SPECTROSCOPY[a]

Wesley W. Wendlandt

Department of Chemistry

University of Houston, Houston, Texas 77004

INTRODUCTION

The measurement of the radiation reflected from a mat surface constitutes the area of spectroscopy known as diffuse reflectance spectroscopy. The reflected radiation may be in the ultraviolet, visible, or infrared regions of the electromagnetic spectrum, although the first two listed are by far the more widely used at present time. The radiation reflected from a mat surface, R_T, consists in general of two components: a _regular reflection_ component (sometimes known as surface- or mirror-reflection), R, and a _diffuse reflection_ component, R_∞ (1,2). The former component is due to the reflection at the surface of single crystallites while the latter arises from the radiation penetrating into the interior of the solid and re-emerging to the surface after being scattered numerous times.

According to the Kubelka and Munk theory (3), the diffuse reflection component, for 1-3 mm thick layers of a powdered sample (an increase in thickness beyond this point has no effect on the reflectance) at a give wavelength is equal to

$$R_\infty = \frac{I}{I_o} = \frac{1 - [k/(k+2s)]^{1/2}}{1 + [k/(k+2s)]^{1/2}} \qquad (1)$$

(a) Research sponsored by AFOSR (SRC)-OAR, U.S.A.F., Grant No. 1190-67.

where I is the reflected radiation, I_o is the incident radiation, k is the absorption coefficient, and s the scattering coefficient. The absorption coefficient is the same as that given by the familiar Beer-Lambert law, $T = e^{-kd}$. The regular reflection component is governed by one of Fresnel's equations

$$R = \frac{I}{I_o} = \frac{(n-1)^2 + n^2K^2}{(n+1)^2 + n^2K^2} \qquad (2)$$

where n is the refractive index, and K is the absorption index, defined through Lambert's law,

$$I = I_o \exp[-4\pi nKd/\lambda_o] \qquad (3)$$

The λ_o denotes the wavelength of the radiation in vacuum and d is the layer thickness.

With some algebraic manipulation, equation (1) can be rewritten into the more familiar form,

$$(1 - R_\infty)^2/2R_\infty = k/s \qquad (4)$$

The left-hand side of the equation is commonly called the <u>remission function</u> or the <u>Kubelka-Munk function</u> and is frequently denoted by $f(R_\infty)$. Experimentally, one seldom measures the absolute diffuse reflecting power of a sample, but rather the relative reflecting power of the sample compared to a suitable white standard. In that case, k = 0 in the spectral region of interest, $R_{\infty std} = 1$ [from equation (4)], and one determines the ratio,

$$\frac{R_{\infty\ sample}}{R_{\infty\ std}} = r_\infty \qquad (5)$$

from which one can determine the ratio, k/s, from the remission function,

$$f(r_\infty) = (1 - r_\infty)^2/2r_\infty = k/s \qquad (6)$$

Taking the logarithm of the remission function gives

$$\log f(r_\infty) = \log k - \log s \qquad (7)$$

Thus, if $\log f(r_\infty)$ is plotted against the wavelength or wave number for a sample, the curve should correspond to the real absorption spectrum of the compound (as determined by transmission measurements) except for the displacement by -log s in the ordinate direction. The curves obtained by such reflectance measurements are generally called characteristic color curves or typical color curves. Sometimes there is a small systematic deviation in the shorter wavelength regions due to the slight increase in the scattering coefficient.

By use of modern double-beam spectrophotometers equipped with some type of a reflectance attachment, r_∞ is automatically plotted against the wavelength. Many investigators replot the data as percent reflectance (%R), or by use of a remission function table (4), plot $f(r_\infty)$ or k/s as a function of wavelength or wavenumber. The most common method is probably the former above.

The above brief introduction to reflectance spectroscopy outlines the most elementary principles of the technique. As would be expected, the technique is widely used for the study of solid or powdered solid samples although it can be used for liquids or paste-like materials as well. The technique is a rapid one for the determination of the "color" of a sample and is generally convenient to use due to readily available commercial instrumentation. Since only the surface of the sample is responsible for the reflection and absorption of the incident radiation, it is widely used in the study of the chemistry and physics of surfaces (5).

HIGH TEMPERATURE REFLECTANCE SPECTROSCOPY

Practically all of the studies in reflectance spectroscopy (it should be noted that the term reflectance spectroscopy used here will denote diffuse reflectance spectroscopy only) have been carried out at ambient temperatures or, in some cases, at subambient temperatures. The latter would most probably be used in single crystal studies for the elucidation of "hot-bands", *i.e.*, transitions which originate from vibrationally excited ground states. However, in many cases, a great amount of additional information on a chemical system can be obtained if the reflectance spectrum of a compound is obtained at elevated temperatures. Normally, temperatures in the range from 100° to 300°C have been used although there is no reason why higher temperatures could not be employed.

Two modes of investigation may be used for high temperature reflectance studies. The first is the measurement of the reflectance spectra of the sample at various fixed or isothermal temperatures. The sample is maintained at a given, pre-determined temperature, and its reflectance spectrum scanned over the wavelength region of interest. The temperature is then changed to a new value and after a suitable equilibration period, the reflectance spectrum again recorded. This procedure may be repeated a number of times until a definite temperature range interval is covered. This mode of operation is called high temperature reflectance spectroscopy (HTRS) and is, of course, a static method in reference to temperature change. The second mode of measurement is to record the reflectance spectrum change of a sample, at a given predetermined wavelength, as a function of temperature. The temperature of the sample is changed (usually increased) at a slow, uniform rate, and the change in reflectance of the sample at some region in the wavelength range is recorded. This mode is called dynamic reflectance spectroscopy (DRS) (9,10) due to the dynamic or changing temperature of the sample.

The above two modes are illustrated by the curves in Figure 1. In the HTRS mode, a series of reflectance curves at various temperatures from 50° to 150° is presented. With increasing temperature, the curve peak at 400 mμ decreases in intensity while the peak at 550 mμ increases. Such a series of curves is possible if the system under study is composed of a single compound which on heating, dissociates to form a new compound having a different reflectance spectrum. By measuring the spectrum of the sample at small temperature increments, the minimum temperature at which the sample begins to undergo the thermal transition can be determined. By use of the dynamic mode (B), the transition temperatures may be determined in a more precise manner. Using a new sample, the change in the reflectance of it (at 400 and 550 mμ) as a function of increasing temperature is recorded. It can easily be seen that the thermal transition begins at about 100° and is completed at about 135°. At 400 mμ, the curve shows a decrease in reflectance as the temperature is increased while at 550 mμ, the reflectance increases over the same temperature range. Although other wavelengths could have been used, the two

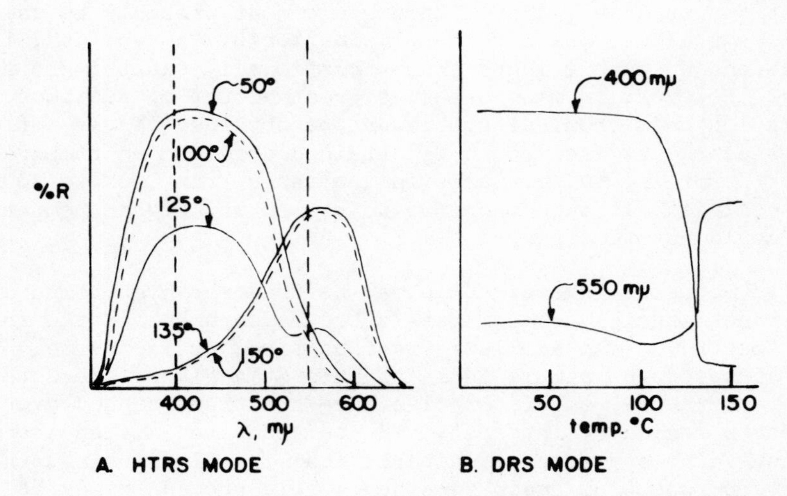

Figure 1. Examples of HTRS and DRS modes.

wavelengths illustrated here indicate the regions of the curve showing the maximum change. The DRS curves (sometimes called isolambdic curves) thus indicate the temperatures at which sample thermal transitions begin and end, and also permit the investigation of only a single thermal transition; mass-loss and enthalpic effects do not interfere with the measurement as is the case with several other thermal techniques. The thermal stability of certain types of compounds can be determined by DRS as well as the elucidation of structural changes. Although many other thermal techniques will yield essentially the same information, DRS is a useful supplementary or complementary technique. Its use in the elucidation of the thermal dissociation reactions and/or reaction mechanisms of coordination compounds have recently been reviewed by Wendlandt (6-8). The technique is also of interest because of its use for the study of absorption, adsorption, oxidation and reduction, and dissociation reactions which may occur on the surface of a compound.

INSTRUMENTATION

The use of a heated sample holder to contain the compound under investigation has been described by several investigators. Asmussen and Andersen (11) studied the reflectance spectra of several $M_2[HgI_4]$ complexes at various elevated temperatures in order to investigate their thermochromic $\beta \to \alpha$ form transitions. The heated sample container consisted of a nickel-plated brass block, 60 mm in diameter by 85 mm in height, the lower end of which contained a chamber in which a small light bulb was mounted. Regulation of the current through the bulb filament permitted temperature regulation of the block. The upper end of the block contained the sample chamber which had the dimensions, 35 mm in diameter by 0.5 mm deep. A copper-Constantan thermocouple, embedded in the powdered sample was used to detect the sample's temperature.

Kortum (2) measured the reflectance spectrum of mercury(II) iodide at 140° but did not describe the heated sample block or other experimental details. Another heated block assembly was described by Hatfield *et al*. (12). It consisted of a metal block into which a heating element was embedded. No other details are available, such as temperature detection, sample thickness, etc.

In 1963, Wendlandt *et al*. (13) described the first of their heated sample holders for high temperature reflectance spectroscopy. The main body of the sample container was 60 mm in diameter by 11 mm thick and was machined from aluminum. The sample itself was contained in a circular indentation, 25 mm in diameter by 1 mm deep, machined on the external face of the cell. Two

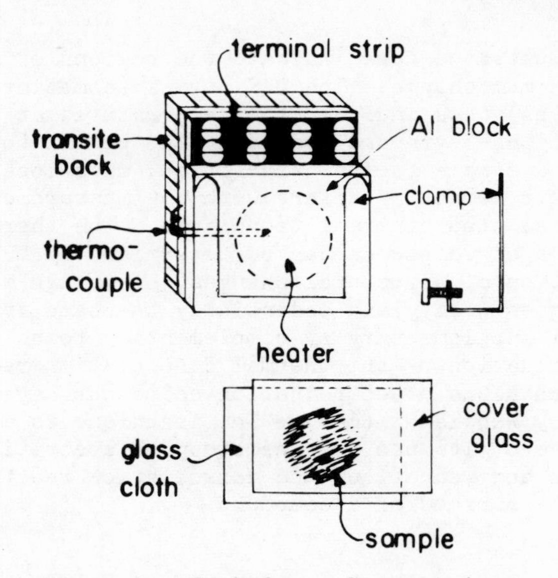

Figure 2. Heated sample container.

circular ridges were cut at regular intervals on the indentation to increase the surface area of the holder and to prevent the compacted powdered sample from falling out of the holder when it was in a vertical position. The sample holder was heated by coils of Nichrome wire wound spirally on an asbestos board and then covered with a thin layer of asbestos paper. Enough wire to provide about 15 ohms of resistance was used. The temperature of the sample was detected by a Chromel-Alumel thermocouple contained in a two-holed ceramic insulator tube. The thermocouple junction made contact with the aluminum block directly behind the sample indentation. To prevent heat transfer from the sample holder to the integrated sphere, a thermal spacer was constructed from a loop of 0.25 in. aluminum tubing and wet shredded asbestos. After drying, the thermal spacer was cemented to the sphere and the sample holder attached to it by a spring-loaded metal clip.

A modification of the above sample holder was described by Wendlandt and George (14) and by Wendlandt (6). The circular aluminum disk of the holder was heated by means of a cartridge heater element inserted directly behind the sample well. Two Chromel-Alumel thermocouples were placed in the block, one adjacent to the heater, the other in the bottom of the sample well so as to be in intimate contact with the compacted sample. The block thermocouple was used to control the temperature programmer while the sample thermocouple was used to detect the sample temperature.

Still another heated sample holder was described by Wendlandt and Hecht (1). It consisted of a block of aluminum, 50 mm in diameter by 25 mm thick, into which was machined a 25 mm by 1 mm deep sample well. A 35 watt stainless-steel sheathed heater cartridge embedded in the main block of the holder was used as the heater. The same two thermocouple system, one for the temperature programmer, the other for sample temperature, was employed. For samples which evolved gaseous products, a Pyrex or quartz cover glass was used to prevent contamination of the integrating sphere.

The latest design of the heated sample holder is shown in Figure 2. This holder is based upon the design of Frei and Frodyma (15) and can be used for very small amounts of sample. The sample is placed as a thin layer on glass fiber cloth which is secured to the heated aluminum metal block by a metal clamp and cover glass. Dimensions of the aluminum block are 4.0 by 5.0 cm. The block is heated by a circular heater element contained within the holder. Electrical connections to the heater and to the thermocouple are made by means of the terminal strip mounted at the top of the assembly. Both the aluminum block and terminal strip are mounted on a 5.0 by 5.0 cm transite block.

The entire DRS apparatus is shown in Figure 3. It consists of an X-Y recorder to record the DRS curve, a strip-chart recorder to record the sample temperature (*versus* time) and a heater controller (16). The reflectance signal for the Y-axis

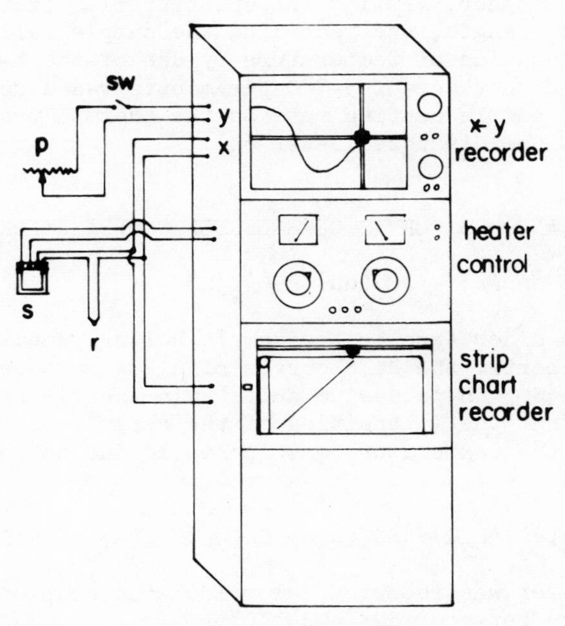

Figure 3. Complete DRS apparatus.

of the DRS curve comes from the spectroreflectometer pen slide-wire, p, while the temperature axis, X, is activated by the thermocouple in the sample holder, S, and reference thermo-junction, r, contained in an ice bath.

The procedure for obtaining a HTRS curve of a sample consists of the following. The sample is finely powdered, using either a manual mortar and pestle or a mechanical grinder (Wig-L-Bug, for example) and placed on the glass fiber cloth in a thin, even layer. The glass fiber cloth is then covered with a 45 mm x 50 mm cover glass and mounted on the aluminum heater block using the metal clamps; the entire assembly is then attached to the integrating sphere of the Beckman DK-2A spectroreflectometer. A room temperature reflectance curve of the sample is first recorded, then the temperature of the block is increased to the desired value and another reflectance curve recorded. This procedure is repeated a number of times until the upper limit of 250° is attained. Recordings are usually made at 25° or 50° intervals but occasionally larger intervals were used. All of the curves were recorded on a single sheet of chart paper using the same wavelength axis. To allow a ready interpretation of the numerous curves, the pen-holder was modified so that Leroy reservoir lettering pens, each containing a different color ink, could be employed.

The DRS curve was obtained by placing a new sample in the heated sample holder, setting the spectroreflectometer to a convenient wavelength, and recording the sample reflectance change as a function of temperature by use of the X-Y recorder. A heating rate increase of 5-7°C per minute was usually employed. The effect of sample heating rate on the thermal transition temperatures will be discussed later on.

THE HTRS AND DRS CURVES OF SELECTED COORDINATION COMPOUNDS

(A) $[Cu(en)(H_2O)_2]SO_4$

Using an older type heated sample holder, Wendlandt (17) previously reported the HTRS curves of a 20% mixture of $[Cu(en)(H_2O)_2]SO_4$ in potassium sulfate in the visible region. He reported that the intensities of the reflectance peak minima increased as the temperature was increased due to the deaquation reaction,

$$[Cu(en)(H_2O)_2]SO_4(s) \rightarrow Cu(en)SO_4(s) + 2H_2O(g)$$

No loss of water was reported below 100° but complete deaquation did take place between 150° and 170°.

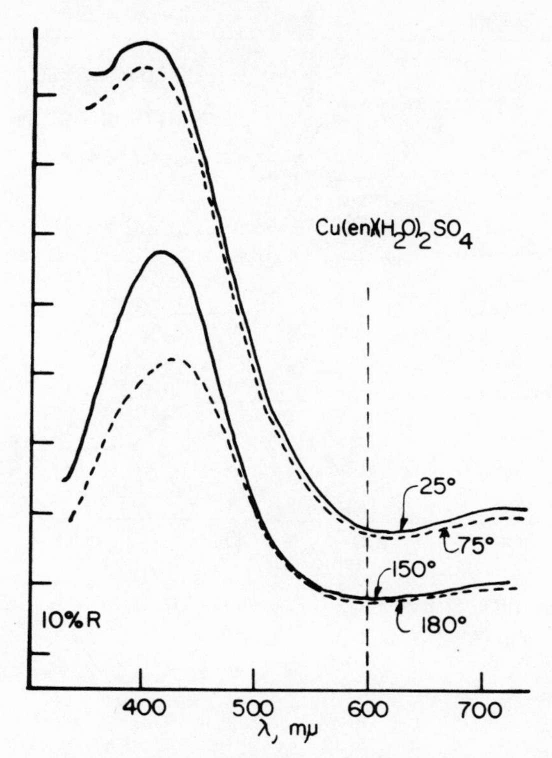

Figure 4. HTRS Curves of $[Cu(en)(H_2O)_2]SO_4$.

Using the new heated sample holder described here, the HTRS curves of $[Cu(en)(H_2O)_2]SO_4$ from 25° to 180° are illustrated in Figure 4. Two sets of curves are shown, one set at 25° and 75°, and the other at 150° and 180°. The first set has a peak minimum at about 625 mμ (corresponding to maximum absorption) and correspond to the curves for the initial compound while the second set as a peak minimum at 575 mμ and correspond tc the curves for the deaquated compound, $Cu(en)SO_4$. Thus, the deaquation reaction must have occurred between 75° and 150°.

To obtain the transition temperature for the deaquation reaction, the DRS technique wa s employed, as shown in Figure 5. The transition temperature dependence upon the sample heating rate is readily seen; it varied from 115° at 6.7°C per min to 165° at 45.8°C per minute. This behavior is not unexpected because it occurs with practically all of the other thermal techniques where some physical property of the sample is measured as a dynamic function of temperature (6). In all DRS studies, the sample heating rate must obviously be specified.

Since the Beckman DK-2A spectroreflectometer is capable of recording the sample spectra in the non-infrared wavelength

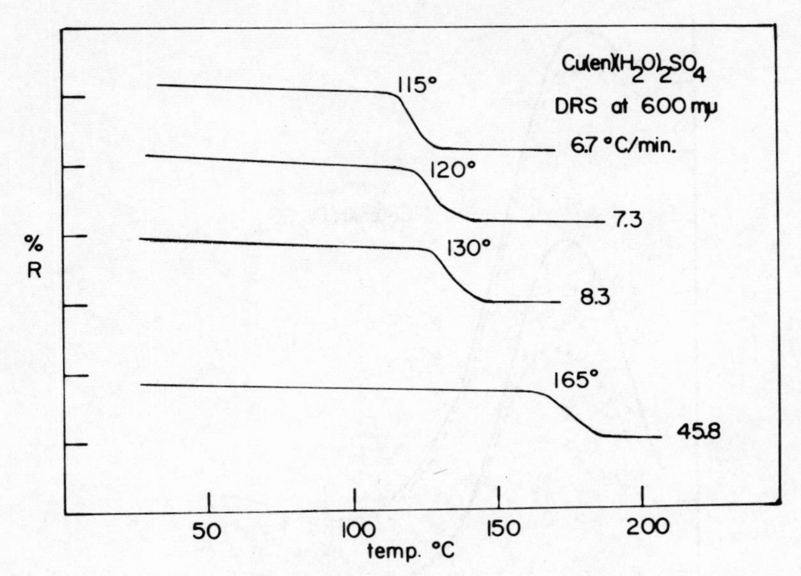

Figure 5. DRS Curves of $[Cu(en)(H_2O)_2]SO_4$ at 600 mμ.

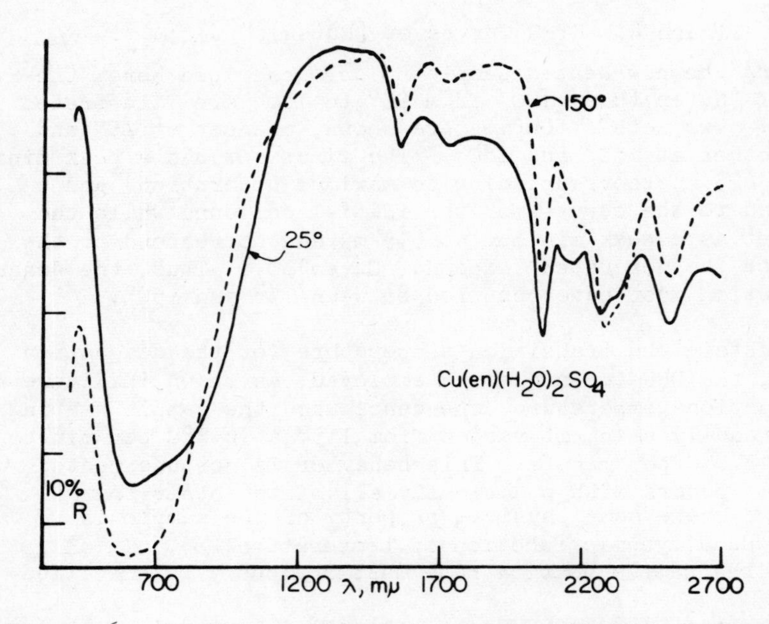

Figure 6. Visible and near-infrared reflectance spectra of $[Cu(en)(H_2O)_2]SO_4$.

region, the HTRS of $[Cu(en)(H_2O)_2]SO_4$ was recorded out to 2700 mμ, as shown in Figure 6. At room temperature, the reflectance curve contained minima (absorption bands) at 1560, 1725, 2050, 2150, 2260, and 2500 mμ, respectively. On heating the sample to 150°, the bands at 1560, 1725, and 2050 mμ remained unchanged while the 2150 mμ disappeared. The 2260 mμ band shifted to 2280 mμ and the 2510 mμ band shifted to 2525 mμ. There were rather pronounced changes in intensity for all of the bands discussed which may be due to the sample particle size changes.

(B) $CuSO_4 \cdot 5H_2O$

The HTRS visible and near-infrared curves are given in Figures 7 and 8, while the DRS curve, at 625 mμ, is given in Figure 9.

As in the case of $[Cu(en)(H_2O)_2]SO_4$, two sets of curves are shown for $CuSO_4 \cdot 5H_2O$ in Figure 7. At room temperature, the

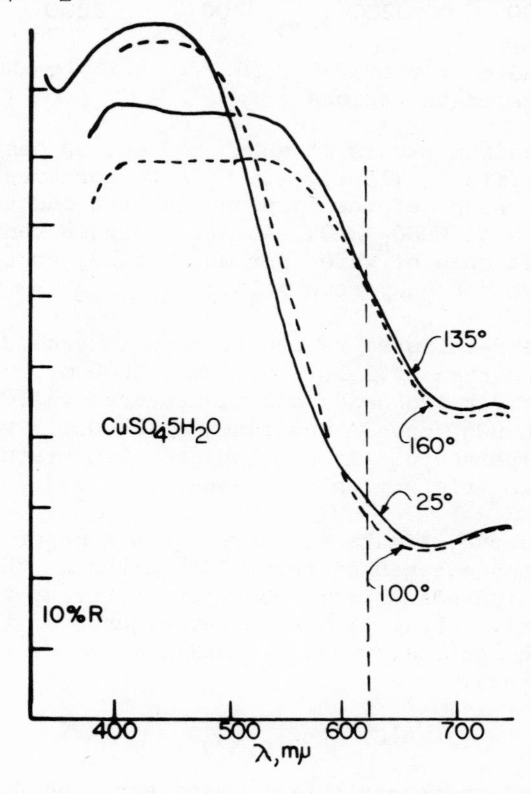

Figure 7. HTRS Curves of $CuSO_4 \cdot 5H_2O$ in the visible wavelength region.

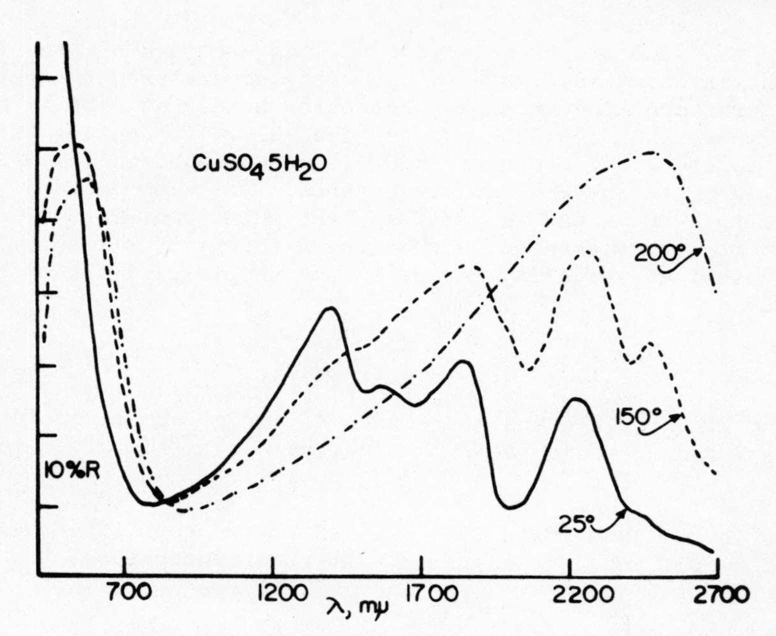

Figure 8. HTRS curves of $CuSO_4 \cdot 5H_2O$ in visible and near-infrared wavelength regions.

reflectance minimum occurs at about 680 mμ; on heating to 135°, the minimum shifts to 715 mμ. In this temperature range, the compositional change of the compound is that due to deaquation from $CuSO_4 \cdot 5H_2O$ to $CuSO_4 \cdot H_2O$. At still higher temperatures, *i.e.*, >250°, the last mole of water per mole of copper sulfate is evolved to give the anhydrous salt.

In the near-infrared region at room temperature, reflectance minima were observed at 1510, 1675, and 2000 mμ, respectively. At 150°, the first two bands had disappeared while the 2000 mμ had shifted to 2060 mμ. A new band, at 2400 mμ, was observed at the higher temperature. At still higher temperatures, 200°, all of the bands in this region were absent.

The DRS curve, Figure 9, showed that a major increase in sample reflectance began at about 105° although the reflectance was gradually increased from room temperature up to 100° (about 0.1 a % R unit). About 125°, the reflectance of the compound again decreased gradually until a maximum temperature of about 200° was attained.

(C) $CoCl_2 \cdot 6H_2O$

In a previous investigation, Wendlandt and Cathers (18) studied the HTRS and DRS of the reaction between $CoCl_2 \cdot 6H_2O$ and

KCl. The DRS curve revealed that the following structural and compositional changes occurred. octahedral- $CoCl_2 \cdot 6H_2O \rightarrow$ tetrahedral- $CoCl_4^{2-} \rightarrow$ octahedral- $CoCl_2 \cdot 2H_2O \rightarrow$ tetrahedral- $CoCl_4^{2-}$. The above reactions took place, of course, in the presence of an excess of chloride ion, hence, the final product was K_2CoCl_4 rather than anhydrous cobalt(II) chloride.

The deaquation of $CoCl_2 \cdot 6H_2O$ was reinvestigated here in the absence of potassium chloride. This compound is a rather difficult one to study because it fuses at about 50° and since the heated sample holder is mounted in a vertical position on the spectroreflectometer, the liquid $CoCl_2 \cdot 6H_2O$ is impossible to retain on the sample holder. This problem was solved, however, by placing a thin layer of the powdered sample on a 25 mm diameter round cover glass which was then retained on the glass fiber cloth by the rectangular cover glass. The viscous nature of the melt prevented the compound from leaving the sample area.

The HTRS and DRS curves of $CoCl_2 \cdot 6H_2O$ are shown in Figures 10 and 11, respectively.

The HTRS curves reveal a rather interesting series of structural changes, both in the liquid and solid states. At 25°, solid $CoCl_2 \cdot 6H_2O$ has an octahedral structure with a reflectance minimum at 535 mμ and shoulder minima at 460 and 500 mμ, respectively. On heating the compound to 55°, it fused and gave a reflectance curve which had one minimum at 525 mμ and a rather broad minimum between 600 and 700 mμ. This latter curve is

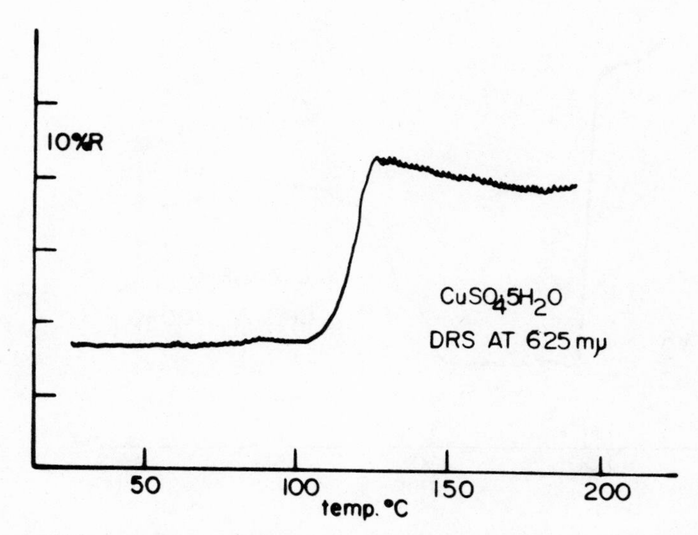

Figure 9. DRS curve of $CuSO_4 \cdot 5H_2O$ at 625 mμ (heating rate 6.7°C per minute.

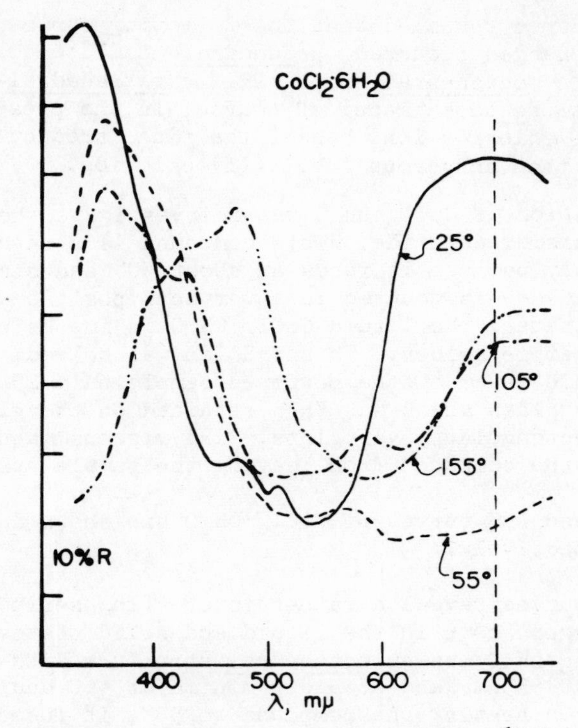

Figure 10. HTRS curves of $CoCl_2 \cdot 6H_2O$.

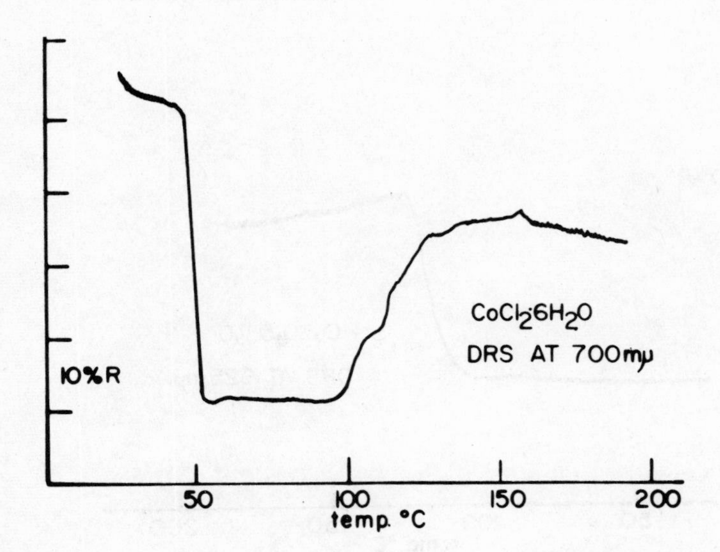

Figure 11. DRS curve of $CoCl_2 \cdot 6H_2O$ at 700 mμ (heating rate of 6.7°C per min.)

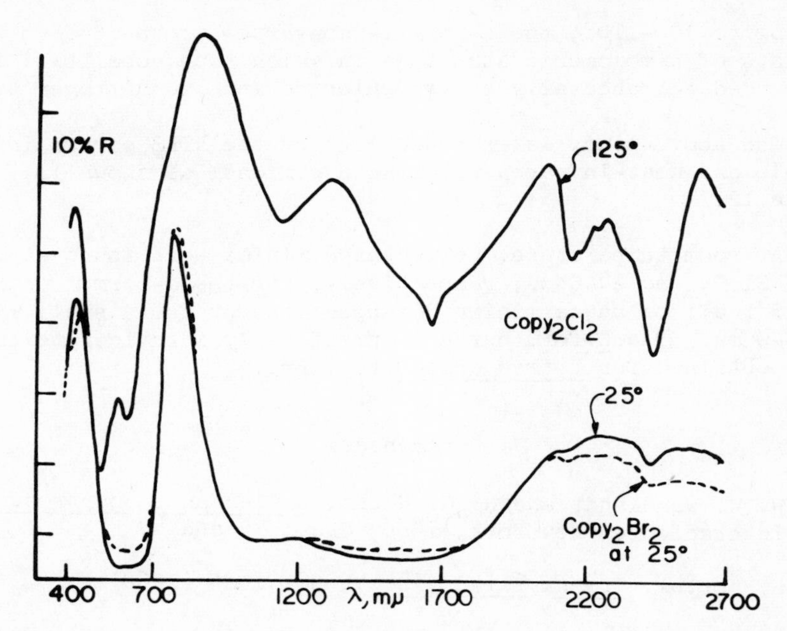

Figure 12. HTRS curves of α-$Co(py)_2Cl_2$ in the visible and near-infrared wavelength regions.

similar to the one previously observed for a mixture of octahedral- and tetrahedral-cobalt(II) complexes by Simmons and Wendlandt (19). Thus, a possible interpretation would be that the 55° curve is probably a mixture of octahedral- $CoCl_2 \cdot 6H_2O$ and tetrahedral-$Co[CoCl_4]$. On further heating, the mixture undergoes further deaquation and gave, at 155°, anhydrous octahedral-$CoCl_2$. This latter curve contained a peak minimum at 590 mμ with a shoulder minimum at 535 mμ.

The DRS curve, Figure 11, showed a pronounced decrease in reflectance at 45° which was due to the formation of the octahedral-tetrahedral mixture. At 100°, the reflectance of the mixture began to increase, reaching a maximum value at about 150°, then decreasing slightly above this temperature. The curve reflects the various structural changes that have previously been discussed.

(D) $Co(py)_2Cl_2$

The HTRS and DRS curves of the α-form of $Co(py)_2Cl_2$ have previously been described by Wendlandt (20). The α-form consists of a polymeric structure in which each cobalt(II) ion is surrounded octahedrally by four chlorine and two nitrogen atoms. On heating to 100-110°, the α-form is converted to the β-form which consists of a monomeric structure in which each cobalt(II) ion is surrounded tetrahedrally by two chlorine and two nitrogen atoms.

The above study is extended here by the HTRS curves in the visible and near-infrared wavelength regions, as shown in Figure 12.

At room temperature, reflectance minima were found at 1140, 1670, 2150, and 2440 mμ, respectively, for the α-form. On heating to 125°, all of these minima disappear except for a small minimum at 2440 mμ. The β-form curve is practically identical to the curve obtained for tetrahedral-$Co(py)_2Br_2$.

References

1. W. W. Wendlandt and H. G. Hecht, Reflectance Spectroscopy, Interscience, New York, 1966, Chap. 3 and 4.

2. G. Kortum, Trans. Faraday Soc., 58, 1624 (1962).

3. P. Kubelka and F. Munk, Z. Techn. Physik, 12, 593 (1931).

4. Reference (1), p. 275-279.

5. Reference (1), Chap. 8.

6. W. W. Wendlandt, Thermal Methods of Analysis, Interscience, New York, 1964, Chap. 10.

7. Reference (1), Chap. 7.

8. W. W. Wendlandt, The Encyclopedia of Chemistry, G. L. Clark and G.G. Hawley, eds., Reinhold, New York, Second Ed., 1966, p. 357.

9. W. W. Wendlandt, Science, 140, 1085 (1963).

10. Anon, Chem. and Eng. News, April 15, 1963, p. 62.

11. R. W. Asmussen and P. Andersen, Acta Chem. Scand., 12, 939 (1958).

12. W. E. Hatfield, T. S. Piper and U. Klabunde, Inorg. Chem., 2, 629 (1963).

13. W. W. Wendlandt, P. H. Franke and J. P. Smith, Anal. Chem., 35, 105 (1963).

14. W. W. Wendlandt and T. D. George, Chemist-Analyst, 53, 100 (1964).

15. R. W. Frei and M. M. Frodyma, Anal. Chim. Acta, 32, 501 (1965).

16. W. W. Wendlandt, J. Chem. Educ., 40, 428 (1963).

17. W. W. Wendlandt, J. Inorg. Nucl. Chem., 25, 833 (1963).

18. W. W. Wendlandt and R. E. Cathers, Chemist-Analyst, 53, 110 (1964).

19. E. L. Simmons and W. W. Wendlandt, J. Inorg. Nucl. Chem., 28, 2187 (1966).

20. W. W. Wendlandt, Chemist-Analyst, 53, 71 (1964).

TECHNIQUES AND ATTACHMENTS FOR A HOHLRAUM-TYPE INFRARED REFLECTANCE SPECTROPHOTOMETER

Roger H. Keith

Research Engineer/Asst. Prof. Chem. Eng.

University of Dayton, Dayton Ohio 45409

The operation of a research infrared spectrophotometer entails inherent problems which arise from the same features which enable the user of such an instrument to perform a wide variety of determinations in a number of modes of operation. Typical difficulties are caused by the complexity of an instrument which must be made to perform a single-beam, dual-beam, programmed, or constant-energy modes with a multitude of source energies and configurations in a number of spectral ranges.

In particular, a hohlraum-type instrument (1) presently implies the use of large optics, long optical paths, and operation of a nickel-oxide-lined cavity in the open air. Previous papers(2, 3) have presented theoretical comment on the operation of a hohlraum reflectance spectrophotometer. This paper is intended more as a practical working guide to time-saving gadgets and versatile methods of operation which permit operation of a hohlraum facility with greater ease, confidence, and versatility. Although most of the techniques and apparatus described were originally intended for use with a Perkin-Elmer Model 13/205 instrument, many can be applied with equal value to other reflectometers. Some techniques which were first applied to transmission instruments are mentioned here because of particular applicability to the problems of hohlraum instruments.

The maintenance and operation of a hohlraum instrument is in many ways like the care of a steam locomotive; a great number of separate subsystems employing diverse media and principles of operation combine to produce the desired result. A variety of skills is needed to attend to the trouble-shooting or modification of any of the working parts. Repairs or additional operating features are often shop-fabricated ad libitum to meet the needs of the moment. Finally, as many complex systems do, the installation exhibits a distinctive anthropomorphic personality which has been described as "one part bathing beauty, one part lumberjack, one part hypochondriac, one part impish small boy."

AMPLIFIER, RECORDER AND DATA REDUCTION

The problem of rapid resetting of a variety of control knobs is easily solved by placing colored dots on the control panel at each of the positions corresponding to normal settings of switches and knobs. A quick glance then suffices to insure that the instrument is ready-to-run in the normal mode of operation after alignment, testing, or trouble-shooting. Batteries used for generation of test voltages can be replaced with zener supplies for worry-free references of infinite life. The normal modus operandi of our instrument implies that the ratio zero control be turned fully off, and a switch to accomplish this was wired around the panel control, eliminating the possibility of a bumped ratio zero control during or before a critical test. Setting of glower current is simplified if the glower supply line is broken at a convenient point and brought to binding posts on the front panel where an accessible ammeter connection may be made. A shorting bar normally preserves the continuity of the circuit across the posts and is removed when the ammeter replaces it.

Recorder operation is simplified if the chart is allowed to dangle rather than being rewound on a takeup spool. A number of clips are available for this purpose, but one can be improvised from a paper clamp and a weight of about 50 grams. A gear box is available (Part No. 124205) from Leeds & Northrup to change the chart drive of the recorder furnished with model 13 instruments so that one wavelength drum turn corresponds to the one-inch chart divisions, making data reduction easier.

A typical chart may contain a 100% line, several sample reflectance lines, and perhaps a zero reference line. A variety of ink colors aids in distinguishing traces, but recorder and chart manufacturers stock an unimaginatively limited palette of hues.

A variety of colors can be compounded using ordinary fountain pen inks, with the addition of water and acetone in various amounts to control intensity of color, surface tension, and drying rate. A bright blue can be formulated from one part Parker Super Quink Washable Royal Blue and one part water. A good green results from the mixture of two parts Parker Super Quink Permanent Green and one part water. Care should be exercised if writing inks are used without dilution; if strong colors such as blue and blue-black are employed, a subsequent trace may wet the original deeper one and cause bleed of colors at the crossover points. A violet can be made from one part Permanent Red and one to two parts Peacock Blue. A rust brown,which is not confused with red, violet, or the commercially-available purples is made from one part Permanent Green and one part Permanent Red. If various colored inks do not provide enough distinctive trace identifications, further diversity can be obtained by lifting the pen frequently to produce a dashed trace.

The long pathlength of the hohlraum beams in open air introduces considerable attenuation due to atmospheric absorption. In addition, many of the mirrors are open to collection of dust, dirt, and degradation products of the hohlraum and its insulation. One consequence of the long pathlength is the difficulty of providing two spectrally identical beams for dual-beam operation, and a 100% line with relaxed flatness tolerances is the result. Often the zero line also exhibits some waviness, and the running of a zero line on each chart is indicated on the most exact work.

Mathematical techniques for handling a continuously varying zero and 100% calibration have been described in the literature[4]. These procedures are simplified if an expanding scale is used to express the reflectance readings. Such scales are made by Gerber Scientific Co. and Ferranti, Ltd., and are employed by placing the zero of the scale upon the zero line at the wavelength of interest and expanding the spring or elastomer scale so that the 100% line corresponds to the 100 graduation of the scale. Intermediate values are read from the stretched scale. A similar procedure can be employed to graduate a chart when three NBS emittance standards[5] are employed: At the wavelength of interest each standard line is aligned with the scale reading corresponding to the NBS value for that specimen at that wavelength. The scale is stretched or translated until the best match is obtained. An indication of system linearity and possible error is obtained by inspection of the closeness of fit which can be obtained. Unknowns are then read from the adjusted scale.

A more versatile 100% line compensator can be used to advantage if the 100% line straightness requires frequent adjustment. The gain of the compensator can be increased by increasing the value of R 439 and decreasing the value of R 426 in the Perkin-Elmer Model 113 amplifier; so long as the sum of these values is 1 kΩ, the correct impedance match will be retained. In our application the ultimate compensation was realized by installing a switch which selected a resistor of 1 kΩ instead of R 439, and shorted across R 426. Compensation at even increments of one drum turn of wavelength was obtained by gearing a ten-turn potentiometer to the wavelength drive with taps spaced at even increments of each drum turn. A built-in set of trimmer pots or a set of interchangeable plug-in pots in banks of six are selectable by panel switches. Trimmers located on each end of the wavelength scan beyond the limits of normal operation permit the adjustment of the slope as well as the position of the 100% line at the limits of traverse. The effect is that the line can be made to go up or down at the end of the chart, and can also be made to do so with increasing or decreasing curvature at will. This adjustment is particularly useful if it is desired to extend the range of the instrument beyond the design spectrum. A wiring diagram of this compensator is presented in the literature (6).

A programmer which scans a drawn curve on a card to produce a continuously varying output as a function of time ("Data-Trak", Research Inc., Minneapolis, Minn.) can be used to generate a voltage proportional to wavelength if the drum-turn vs. wavelength calibration is drawn upon the program card. The output of the card programmer can be fed to the x-axis and the output of the spectrophotometer can be fed to the y-axis of an x-y recorder to produce a chart having a linear wavelength scale.

Slotted box conduit aids immeasurably in presenting a neat, orderly arrangement of cables, wires, hoses, and pipes needed for the operation of a hohlraum spectrophotometer. Plastic conduit rack (Panduit Corp., Tinley Park, Illinois) can be cut and drilled easily for this purpose.

THE OPTICAL PATH

Optical alignment is often critical, due to the long path length, exposure of the optical elements to air and dust, thermal expansions induced by the hohlraum furnace, and the strict performance requirements of research calibration determinations. Many hohlraum-instrument owners prefer to perform their own frequent optical alignments as a result, and it is common to find a hot-cavity operator who "rolls his own."

The mounting of the instrument can be of importance. A heavy welded framework can provide a sturdy base which will decrease the warping and disalignment of the instrument. If the frame does not have a plane top, but rather a ledge around the outside on which the instrument feet rest, then it becomes possible to work upon the underside of the spectrophotometer without disturbing its position. Vibration-damping pads under the framework feet aid materially in reducing the effect of seismic disturbances upon the instrument. "Visorb" jack feet (Enterprise Machine Parts, Detroit, Michigan) provide this isolation along with a levelling adjustment.

A convenient mercury-arc lamp holder to facilitate frequent optical alignment of hohlraum spectrophotometers is dimensioned in the literature [7]. An earlier article by Visapää[8] describes the principle of operation. Recent experiments in aligning the monochromator with a CW gas laser have been eminently successful. Our laboratory has found that all but the finest slit adjustments can be made in ordinary room light. A laser is also useful to project a beam backwards from the thermocouple station through the entire instrument to check sample positioning in the hohlraum. The projected slit image is visible on the sample even when the hohlraum is at operating temperature.

Other adapters can be made which slip into the transmission sample holders of the instrument. It has been found that adapters to accept 2x2 glass color filters or Series VI photographic filters are useful in instrument wavelength calibration or stray-light filtering.

Many materials may be employed in wavelength calibration. Glass color filters provide a convenient standard but have little or no pass beyond 5 microns. Didymium and holmium oxide glass color filters (CS 1-60, CS-3-138, Corning Glass Works, Corning, New York) have band structures useful for spectrometer calibration.

Filters CS 9-54, 4-77, and 4-106 also exhibit bands in the 1 to 5 micron region which are suitable for rough calibrations. The mercury arc is also a useful source of IR bands which may be employed in coarse work [9].

Conventional high-accuracy wavelength standard materials are described by a number of authors [10, 11, 12]. Indene has been found to be a useful one-material standard with a large number of bands throughout the 2 to 15 micron region. Band coverage is enhanced if cyclohexane and camphor are added to the indene [13].

Alignments are facilitated if a set of jigs for the 205 reflectance attachment are fabricated. These jigs should be analogous to those described in the Model 13 instrument manual [14] with dimensions suitably altered to place wires, crosses, etc. at the centers of the optic axes. An auxilary area lighting system with red bulbs is time-saving when green-lining the instrument with mercury light. The red general illumination does not interfere with the visibility of the beams, yet misplaced tools or inaccessible adjustments are easily located. It would seem equally feasable to use green illumination if a neon laser were employed in alignment.

Plastic sandwich bags can be used advantageously to cover parts of the instrument when not in use, or to protect them from humidity or fingerprints during alignment. A durable sheet of cardboard, asbestos board, or aluminum cut to fit over the hohlraum foldup mirrors can avert catastrophe if water begins to cascade from a leaky specimen through the incandescent hohlraum onto the hot front-surface optics. (Our laboratory has permanently removed the sheet metal covers from the hohlraum optical section to facilitate such emergency tactics, and to eliminate temperature changes -- and resulting optical mount warping -- when the covers are replaced following alignment.) It is inevitable that particles of the hohlraum insulation will decrepitate and fall onto the optics below. These grit particles may be partially removed by blowing with a infant's ear syringe. It is not recommended that a brush be used; even an anti-static radioactive camel-hair lens brush soon becomes loaded with abrasive grit and scratches the mirror surface. The technique of coating the surface with flexible collodion and stripping of the dried coat along with the accumulated dirt[15] has been found to be most effective and has been adopted as standard practice.

A sheet metal guard is placed over the emittance furnace diagonal to protect it from accumulated dust when not actually in use. Another sheet metal guard is placed under the mirror to protect its front surface from possible vapor deposits from the heated emittance sample during its warmup period; this guard is removed only while data is actually being taken.

A transparent plastic box has been fabricated to fit over the transmittance sample section of the spectrophotometer, and vinyl plastic bellows (Gagne Associates, Binghamton, N. Y.) can be applied to the path between the transfer section and the monochromator, permitting inerting of the model 13 section of the instrument without fitting windows. (These covers find their best use in preventing the accumulation of dust and dirt, on the optics, however, rather than as inerting envelopes.)

A No. 3 fruit juice can has been adapted to use as an inerting enclosure for easily-oxidizable specimens placed in the emitting furnace of the 205 reflectance/emittance attachment. A salt window is placed in the path of the existing beam and the can rests upon an asbestos-cloth gasket. Details are given in a previous report[16].

An interrupter for the beams of the spectrophotometer is a useful device for evaluation of the energy available for measurement, and thus gives an indication of accuracy. A simple device for this purpose is described by Carrington[17] for use when solvents are used in the reference beam of a transmission spectrometer. A similar situation exists with atmospheric absorption in a hohlraum reflectance instrument and the interrupter is a valuable tool.

The breaker assembly, which de-commutates the signals from the amplifier corresponding to the optical chopper position, is required to operate reliably for long periods of time in a difficult environment. In addition to operating one million times in the course of a day's work, the chopper and breaker assembly must withstand dust and dirt since the rear of the chopper section housing is not equipped with salt sindows. (Many users of the apparatus run entirely without windows, since the pathlength which can be inerted through their use is short compared with the open paths in the reflectance attachment, and the window losses are thus eliminated.) The heat of the nearby hohlraum adds an operational difficulty.

Electrical connections to the chopper have been patched at the terminal board inside the chopper housing. Switches and jack sockets exterior to the instrument permit monitoring of voltage or current waveforms at any terminals with an oscilloscope while the instrument is in operation with covers in place. The same auxiliary panel holds batteries, lamps, and switches for performing breaker-vs.-chopper alignments described in the instrument manual merely by throwing switches.

In operation, the chopper shaft typically develops a 13cps squeak which is either adapted to by the operator, or becomes maddening within minutes. The use of ordinary oils is advised against because of their capillarity which causes them to spread to the breaker contacts, and because of vapors which can deposit on mirrors. It has been found that dolphin oils possess the required thixotropic rheology and in the form of clock oil (Wm. F. Nye, New Bedford, Mass.) have been used with success for years on our instrument bearings and cams.

Another approach has been taken by the Thermophysical Properties Lab at Wright-Patterson AFB, who have replaced the bronze bushings of the chopper shaft with ball bearings. In 1962, our laboratory made preliminary experiments with a solid switch (GE ZJ 235) to activate a controlled rectifier, using contactless photoelectric sensing of the chopper position to drive the switch. This device has breadboarded [18] but never placed in operation because of doubts that the reliability of the factory design could be exceeded. Replacement choppers are now commercially available which cover a variety of speeds (Brower Laboratories, Inc., Westboro, Mass.)

The sample beam final hohlraum foldup beam mirror [19] is mounted kinematically so it may be removed to allow the energy to come from the emittance sample when running in the emittance mode. Addition of a kinematic mount to the foldup mirror in the reference beam allows checks on the beam balance in the emittance mode and permits easy access to both beams, facilitating more versatile operation of the instrument as a research photometer.

In the emittance mode there is no way a 100% line can be run to check whether either beam has suffered a degradation of energy due to mirror aging, improper alignment, or beam blockage. By supplying each foldup mirror with a kinematic mount, a reference sample may be checked, first with the emitted energy in the sample beam, and then in the reference beam, by removing one or the other of the kinematically-mounted mirrors. (The emittance furnace can be aligned with the reference beam by removing the asbestos-board top plate of the furnace and turning it over. Upon replacement in the reversed postion, the board will have flipped into alignemnt with the rear beam.) It should be noted that even with kinematic mounts on both beams, although the I_0 mirror train can be checked against the I mirror train, in each case the beam from the sample passes over the sample diagonal mirror, whereas the beam from the hohlraum passes over the foldup diagonal, and full reversal of beam symmetry is not obtained. It should also be borne in mind that although the measured pathlengths to the emittance sample and hohlraum are the same as determined by focus alignment at room temperature, when the hohlraum is at operating temperature the reference pathlength is optically shortened due to the expansion of the hohlraum gas at elevated temperatures. The emittance sample causes heating of the air above it, also, but the temperature is not nearly so high nor is the pathlength so long as in the hohlraum.

If the instrument is used as a photometer to compare energies from a source at temperatures above 1100°C, an attenuator to reduce the sample beam energy to the levels emitted by the hohlraum is needed, so that recorder readings will be below 100% and within the linear range of the instrument. Large attenuators with flat transmission characteristics can be easily fabricated by parallelogramming ordinary window screen to produce the desired open area. Transmittances on the order of 40 to 60% can be produced in this manner, and large screens which are quite even over their areas can be easily made to fit the oversized beams of the hohlraum section. The screens are tacked down to a board in the desired orientation and then permanently bonded with solder, capacitance welding, or gluing to a frame. Transmittances below 40% are obtained by sandwiching two or three layers of parallelogrammed screening and joining them permanently. Transmittances as low as 10% are thus obtained. Careful choice of the degree of parallelogramming and the orientation of the layers will prevent gross moiré patterning. The screens are simply calibrated by running instrument 100% lines with and without screens in place.

Specular reflectance measurements may be made with the 205 attachment by mounting a 1x1 inch specimen in place of the square mirror of the carriage which fits into the Model 13 sample space. This mirror was selected for replacement since it is easily adjusted, but any other small diagonal would serve as well. The mirror is adjusted for maximum energy and the specular reflectance of the specimen is obtained by running the instrument as in the 100% line determination. The USAF Thermophysical Properties lab at WPAFB has constructed a special mirror carriage for this purpose, incorporating calibrated vernier mirror angular adjustment controls.

HOHLRAUM

Two-position control has been substituted for the on-off control originally supplied for the hohlraum. This mode of control provides even greater smoothness of temperature with little cost. The original thermocouples and controller have been retained, but the controller now selects either a high or a low voltage (the lower one being set to maintain a temperature just below the desired set point and the higher voltage corresponding to a temperature just above the set point.) Continuously variable voltages in a fixed ratio are obtained by selecting either the 0-110 or 0-140 volt ranges of a variable autotransformer. An autotransformer knob setting of 75 on the 140 volt scale suffices for operation at 1000°C.

In our laboratory, the hohlraum has been run continuously at high temperature for a number of years, with the benefits that little or no spalling of the hohlraum oxide liner and very little sifting of the furnace insulation occurs. The hohlraum is normally operated at 1000°C for reflectance determinations, and is depressed to 900°C when not in use. The cavity can be returned to operating temperature within 30 minutes.

A cap is placed on the sample holder tube of the hohlraum to prevent excessive thermal losses when the sample holder is not in place. A steel disc about four inches in diameter with a depressed center to act as a "hat with a brim" is employed. Cigarette companies often give away pressed steel ash trays which give admirable service.

A thermocouple probe for checking hohlraum internal temperatures is a necessary accessory. A thermocouple is mounted on spacers made so that it is located at the sample focus in use. A cap at the top of the sample holder tube stops heat losses. Frequent checks of hohlraum operation are mandatory for well-defined reflectance or emittance measurements; control thermocouples mounted outside the hohlraum often show errors in excess of 20°C.

The hohlraum temperature-check thermocouple probe can be calibrated vs. an NBS thermocouple, or can be viewed with a standardized optical pyrometer. This may be done by tilting one of the hohlraum foldup mirrors so that a clear view of the cavity interior is obtained with the pyrometer. The effect of the imperfect reflectance of the mirror which is interspersed in the optical path is the same as the effect of imperfect target emittance and the same corrections can be employed. (The reflectance of the mirror can be ascertained by sighting on an incandescent body directly with the pyrometer and comparing the reading with that obtained by use of the foldup mirror at the same angle employed in the viewing of the hohlraum.)

A simpler -- but less comfortable -- method of checking thermocouple temperatures with an optical pyrometer results if one bores a one-inch hole through the baseplate of the instrument directly beneath the hohlraum. One can then sight directly upwards into the hohlraum from beneath the instrument. The proper location for the best view can be determined by inspecting the shadows of the hohlraum mirrors on the baseplate and choosing the largest sunny area as the place to begin drilling operations.

This modification allows use of the hohlraum as a precise calibrating blackbody for optical pyrometers, thermocouples, and other types of thermal sensors in the 20-1100° range.

A motor-driven furnace rotator to position the hohlraum reproducibly in the 100% and sample positions is an aid to accurate results as well as a convenience to the operator. Electrical limit switches provide positive location and do not introduce hohlraum tilt as the originally-furnished locking device can do if carelessly operated. A slow-speed reversible motor drives a toothed rack which is bent to conform to the surface of the furnace exterior, through an idler gear. Limit switch stops are attached to steel banding straps which girdle the furnace.

SAMPLE COOLING

The reflectance sample is immersed in a blackbody furnace at high temperature in hohlraum-source testing, and drastic measures must be taken to insure that the specimen remains cool. If specimen's temperature rises appreciably it will emit radiation itself and cause abnormally high reflectance readings. A concomitant problem is the thermal degradation or outright destruction of a thermally-sensitive material.

A more highly instrumented coolant control panel has been constructed to provide closer control and better regulation of sample coolant characteristics. The panel features gages to monitor hot and cold water inlet pressures, large dial thermometers on inlet and exit to both sample and jacket chambers, micrometer needle valves to control sample and jacket flows, low pressure and low flow alarms, and rotameters to indicate jacket and sample flow rates (Safeguard 18210, 250mm, 3RB-3LA, Schutte and Koerting, Cornwells Hts., Pa.) A cable of four 3/8 inch tubes (Okoflex 782-204, Okonite Co., Passaic, N. J.) replaces the somewhat unwieldy bundle of separate tubes to the sample holder; the larger diameter of the tubing provides higher flow rates and better cooling. Furnace thermocouples are also bound in a multi-strand cable for neatness. (PPK 6-16-CL, Thermo-Electric, Inc., Saddle Brook, N. J.)

Compressed air is automatically admitted to the sample holder by the coolant control when coolant water is shut off, blowing out the sample from the rear and permitting disassembly of the sample holder with less untidyness and less chance of ruining a water-sensitive coating.

Silicone Rubber (CS-1501, 0.061 inch, Anchor Rubber Co., Dayton, Ohio) has been found to be an excellent material for sample holder gaskets, possessing the combination of high temperature resistance, resilience, and simple fabrication. A cookie-cutter die suffices for forming the gaskets, although a dual die incorporating concentric cutting surfaces makes gasket manufacture less frustrating.

A small telescope has been fitted to the Reflectance Attachment frame to permit viewing the specimen in situ during operation of the spectrophotometer. This telescope (18 E 2193 Finder, 7X, A. Jaegers, Lynbrook, N. Y.) is fitted with a photographic +1 portrait lens and views the specimen through a diagonal mirror mounted between the hohlraum foldup mirrors. The telescope is invaluable in assuring reproducible alignment of the tilted hemispherical reflectance cap with the sample beam axis, determining alignments of optic axes during instrument green-lining, examining the specimen under operating conditions to insure that a good gasket seal is maintained, and to provide a visual check on possible degradation of a sample on exposure to hohlraum radiation.

SAMPLE TECHNIQUES

Sample substrates may be fabricated by a number of methods. Machinining is perhaps the most expensive, but provides the ultimate in control of surface finish, flatness, and dimensional tolerances. Techniques for punching specimens with excellent control have been developed to a high degree[20]. A hand punch (No. 118, Whitney Jensen, Rockford, Ill.) forms discs from aluminum sheet with excellent tolerances and freedom from burrs. The sheet is layered with lens tissue to prevent surface scuffing and compressed between bristol sheets before punching. The protective laminations reduce burring and dishing of the finished pieces. The dishing is particularly undesirable in specular-surfaced samples since it can cause the specimen to act as a relay lens to refocus a small part of the hohlraum (which may be an anomalous temperature) over the entire specimen surface.

Soft coatings may be removed from the edges of post-coated substrates by carving or filing while the specimen is held by the edges. Vitreous coatings may be removed from the substrate edges by filing downward over the disc edge.

Thin specimens may be run by shimming with aluminum washers behind the specimen to build an aggregate thickness of 0.035 inches or greater. Aluminum samples as thin as 0.008 inches have been successfully tested in this manner, and did not appear to dish due to cooling water pressure. The ultimate lower limit to the thickness of samples which can be accomodated using shim washers seems to be set by handling or shipping difficulties with fragile specimens rather than any structural weakness during test.

Specimens thicker than 0.065 inches could conceivably be handled by modifying the sample holder cap or by lapping the sample to a lesser thickness. The first method is to be discouraged because of the heat transfer difficulties attendant to thick samples and the second method risks destruction or subtle change of surface properties during the lapping process. At the present time there seems to be no simple solution to the thick-sample problem aside from outright refusal to run oversize substrates.

Organic thin films with high reflectances, which thus have relatively low heat absorption and low cooling requirements, can be bonded to metallic substrates for determination of reflectance[21]. An epoxy cement has been used for this purpose, but a simpler method employs an adhesive transfer tape (No. 465, 3M Corp., St. Paul, Minn.) which is burnished onto the substrate. The tape backing is peeled off, with the adhesive layer remaining on the substrate. The thin film to be tested is then applied to the adhesive, covered with protective tissue, and pressed or burnished to set the adhesive bond. In employment of this method, it should be borne in mind that the substrate and adhesive will "show through" and exert an influence on results if the film is not opaque.

Small specimens can be run by masking down the entrance slits of the monochromator so that only a small portion of the normal sample is viewed and is compared with an equivalent area of hohlraum wall in the double-beam mode. A practical limit of about 6 or 7 mm as the smallest dimension of the sample is set by the instrument aberrations resulting from an entrance-slit mask of 0.5 x 0.5 mm. and the masking of the sample edge by a lip or angle of the cap used to clamp the sample to the holder.

A number of sample caps cut out to fit small rectangular specimens have been fabricated in our laboratory, along with specially cut water-sealing washers to match. Each of these adapter caps has been intricate to design, expensive to make, unforgiving of poor tolerances, tedious to assemble, and almost impossible to seal into a leakproof unit.

On the contrary, a design by Blair of Hughes Aircraft[20] has allowed the determination of reflectance of 7/16 x 15/16 inch rectangular samples of varying thicknesses with uneventful reliability. This adapter relies on a sample-disc-like insert which provides small-sample capability with conventional washers and sample cap, rather than with a specially-designed cap and washer. The adapter in its original form is a copper disc, 2.5 mm thick, 25.4 mm in diameter, with a 1.5 mm deep x 12 mm wide slit milled in one side along a diameter. As adapted in our laboratory, the disc is slightly smaller in diameter to fit the tilted hemispherical reflectance sample holder cap, and is 24K gold plated. In both the original and modified discs, the rectangular sample extends to the edge of the adapter disc, riding in the disc grove. The sample is clamped to the disc by the lip of the conventional capping ring, which nips the corners of the small rectangular specimen. The specimen is aligned in the hohlraum so that the long axis of the specimen coincides with the monochromator slits.

In the modified adapter disc, the depth of the milled grove was reduced to about 1 mm, so that thinner specimens would still extend above the plane surface of the disc and be exposed to strong clamping action from the cap lip. Preliminary trials with this adapter were made using transistor heat sink compound (Type 340, Dow-Corning Corp., Midland, Mich.) to provide thermal coupling between the sample and the adapter disc, but tests disclosed this to be an unnecessary precaution and possible source of sample contamination. A comparison between a black 15/16 inch conventional reflectance disc and a small rectangular specimen coated with the same black (Krylon No. 1602, Borden Chem. Co., New York, N.Y.) ungreased and clamped to the gold adapter disc, showed no measurable differences in charted reflectance between 1.5 and 15 microns. Sample and jacket flows were each 0.7 gpm at 60°F and the hohlraum was at 1000°C., the usual operating conditions at this laboratory. The results of this test showed that even under these extreme conditions of heat absorption and contrast between the sample and the adapter disc, thermal contact without grease was sufficiently effective to prevent specimen-emittance dilution of data, and the monochromator field of view with an unmasked slit does not spill over onto the adapter disc.

Several ingenious methods of shipping or storing samples have been devised by various laboratories. We have found that the cardboard booklets used by coin collectors for displaying U.S. Quarter collections are ideal for the ordinary reflectance specimens, (No. 9044, Whitman Pub. Co., Racine, Wis.) which fit the apertures loosely but do not rattle. Delicate surfaces may be protected by layering lens tissue or plastic films over the filled cards. Sample designations may be lettered below the filled spaces to identify the discs. Square 2x2 inch specimens employed in other reflectance spectrophotometers are easily stored or shipped in cases or slide trays intended for 35 mm slides. Envelopes of foamed polyethylene (Foamvelopes, Donray Products, Cleveland, Ohio) can be used as an inexpensive package for intermediate wrapping of fragile specimens for shipping or storage.

ACKNOWLEDGEMENT

The author wishes to express his appreciation to Robert Anacreon and Melville Nathanson of Perkin-Elmer for their guidance and patience, to many personnel of the USAF Materials Lab Coatings Branch for their support, suggestions, and understanding; to all those who contributed their ideas and efforts to this work, and in particular to many at the U. of Dayton Research Institute who have given unselfishly of their time and talent.

A research spectrophotometer and its accessories are much like a sewing machine which comes with a box containing a button-holer, a pleater attachment, a zipper foot, and so forth. Although the sewing machine is acquired principally for stitching in straight lines, all the fun lies in using the attachments. In the same way, much of the fun of a hohlraum spectrophotometer seems to be in making it do something it was never intended to do, and doing it better than can be done in any other way.

REFERENCES

(1) J. T. Gier, R V. Dunkle, and J. T. Bevans. J. Opt. Soc. Am. 44:558.

(2) W. L. Starr, and E. Streed J. Opt. Soc. Am. 45:584.

(3) R. J. Hembach, L. Hemmerdinger, and A. J. Katz, in "Measurement of Thermal Radiation Properties of Solids" (J. C. Richmond, ed.) pp. 153-167, Ofc. Sci. Techn. Inf., NASA SP-31, 1963.

(4) E. R. Streed, L. A McKellar, R. Rolling, Jr., and C.A.Smith, in "Measurement of Thermal Radiation Properties of Solids" (J. C. Richmond, ed.) pp. 237-252 Ofc. Sci. & Techn. Inf., NASA SP-31, 1963.

(5) Anon. NBS Technical News Bulletin, 1963:185.

(6) R. H. Keith, "Attachments for a Recording Spectrophotometer," U. Dayton Res. Inst., Dayton, Ohio. in press.

(7) R. H. Keith, Appl. Opt., 5:1334.

(8) A. Visapää, Perkin-Elmer Instrument News, 13 (3): 12.

(9) W. Uyterhoeven, "Elektrische Gasentladungslampen" Springer, Berlin, 1938, quoted in V. K. Zworykin & E. G. Ramberg, "Photoelectricity & It's Application", p. 21, Wiley, 1961.

(10) A. Danti, L. R. Blaine, and E. D. Tidwell. NBS J. Res. 64:29.

(11) G. K. T. Conn and D. G. Avery, "Infrared Methods", p. 171, Academic Press, 1960.

(12) Anon. "Instruction Manual, Model 13 Ratio Recording IR Spectrophotometer Operating and Maintenance Instructions" Vol 3 C, pp. 31-36, Perkin-Elmer Corp., Publ. 90-9004, Norwalk, Conn. 1955.

(13) H. N. Jones, N. B. W. Jonathan, M. A. MacKenzie and A. Nadeau. Spectrochim. Acta, 17:77.

(14) Anon. "Model 13-U Alignment and Test Procedures", 15pp. Instrument Div., Training Sect. Perkin-Elmer Corp., Norwalk, Conn. (1960).

(15) J. B. McDaniel, Appl. Opt., 3:152.

(16) R. H. Keith "The Preparation of Samples for Determination of Their IR Optical Properties using the P-E 13/205 Apparatus." UDRI-TR 64-105, U. Dayton Res. Inst., Dayton, Ohio, (1964).

(17) R. A. G. Carrington, J. Sci Instrum. 43:75.

(18) K. Perko, "Perkin-Elmer Model 13 IR Spectrophotometer with Model 205 Reflectance/Emittance Attachment" paper submitted to EE Dept. Seminars, U. Dayton, Dayton, Ohio 1962.

(19) Anon, "Infrared Reflectivity Attachments" Publ. 990-9188, Perkin-Elmer Corp., Norwalk, Conn.

(20) C. Boebel, H. H. Hormann, J. Mattice, G. Stevenson, and J. Weaver, private communications, Coatings Branch, USAF Materials Lab, Wright-Patterson AFB, Ohio, 1962-67.

(21) G. Goken, private communication, 3M Corp., St. Paul Minn., 1967.

ANALYSIS BY MEANS OF SPECTRAL REFLECTANCE OF SUBSTANCES RESOLVED ON THIN PLATES

Michael M. Frodyma[(1)] and Van T. Lieu[(2)]

(1) National Science Foundation, Washington, D. C.

(2) Dept. of Chemistry, California State College, Long Beach, California

INTRODUCTION

Because of the many advantages it affords, thin-layer chromatography has found widespread application in the resolution of mixtures and the identification of their components. Not only are the procedures employed simple and easily carried out and the equipment required inexpensive, but the separations achieved are usually sharper than those obtained with the same or a similar solvent-system used in conjunction with paper. Particularly noteworthy is the speed of chromatographic development, with only one or two hours often sufficing for the resolution of a mixture on chromatoplates.

Despite these advantages, however, the analytical utility of the technique is restricted by two shortcomings. In the first place, the difficulty experienced in obtaining reproducible R_f values with thin-plates frequently makes it necessary to run standards alongside the samples for comparison purposes. Secondly, the quantitative removal and extraction of individual spots is a tedious process which often cannot be accomplished without losses occurring. Both of these operations would become superfluous if it were possible to effect the *in situ* identification and determination of chemical species separated on thin-plates.

The use of spectral reflectance for these purposes was suggested by various studies[(1-4)] which demonstrated its utility with respect to paper chromatography. Furthermore, it has been shown that the reflectance spectra of substances concentrated on particulate adsorbents can be used for their identification,[(5)] and that spectral reflectance can be employed to determine the concentration of dyes scavenged from solution by the batchwise addition of starch.[6] A critical evaluation of the application of reflectance measurements to the direct analysis of solid mixtures has established the fact that analytically useful data can usually be obtained with samples in powdered form.[7] In view of these results it was decided to investigate the analytical applications of spectral reflectance to thin-layer chromatography.

THEORY

Kubelka-Munk Equation

For the purpose at hand probably the most appropriate theory treating diffuse reflection and the transmission of light-scattering layers is the general theory developed by Kubelka and Munk.[8-9] For an infinitely thick, opaque layer (achieved in practice with a layer thickness of a few millimeters) the Kubelka-Munk equation may be written as

$$F(R_\infty) \equiv (1-R_\infty)^2/2R_\infty = k/s \qquad (1)$$

where R_∞ is the diffuse reflectance of the layer relative to a nonabsorbing standard such as magnesium oxide, k is the molar absorption coefficient of the sample, and s is the scattering coefficient. Provided s remains constant, a linear relationship should be observed between $F(R_\infty)$ and k. This has been confirmed for weakly absorbing materials where the contribution of regular reflection is small.[10]

When the reflectance of a sample diluted with a non- or low-absorbing powder is measured against the pure powder, the absorption coefficient, k, may be replaced by the product $2.303\,\varepsilon c$, where ε is the extinction coefficient and c is the molar concentration.[11] The Kubelka-Munk equation, (1), can then be written in the form

$$F(R_\infty) \equiv (1-R_\infty)^2/2R_\infty = c/k' \qquad (2)$$

where k' is a constant equal to $s/2.303\varepsilon$. Since $F(R_\infty)$ is proportional to the molar concentration under constant experimental conditions, the Kubelka-Munk relationship is analogous to the Beer-Lambert law of absorption spectrophotometry. At high enough dilutions, the regular reflection from the sample approximates that from the standard and is thus cancelled out in any comparison measurement.

It has been found that the common, commercially available thin-layer adsorbents are suitable for use as diluents in both the visible and ultraviolet regions of the spectrum. When such adsorbents are mixed in a small mortar with appropriate amounts of the substance undergoing analysis, the resulting system usually conforms to the Kubelka-Munk equation over a concentration range that has analytical utility.[12,13]

Optimum Concentration Range and Preparation of Calibration Curves for Reflectance Analysis

For systems exhibiting no deviation from the Kubelka-Munk equation, the optimum conditions for maximum accuracy can be deduced by computing the relative error, dc/c. In terms of the Kubelka-Munk equation, (2), the error in c is

$$dc = k'(R_\infty^2 - 1)dR_\infty / 2R_\infty^2 , \quad (3)$$

and the relative error in c is

$$dc/c = (R_\infty + 1)dR_\infty / (R_\infty - 1)R_\infty . \quad (4)$$

Assuming a reading error amounting to one reflectance unit, that is $dR_\infty = 0.01$,

$$(dc/c) \times 100 = (R_\infty + 1)/(R_\infty - 1)R_\infty = \% \text{ error in c}. \quad (5)$$

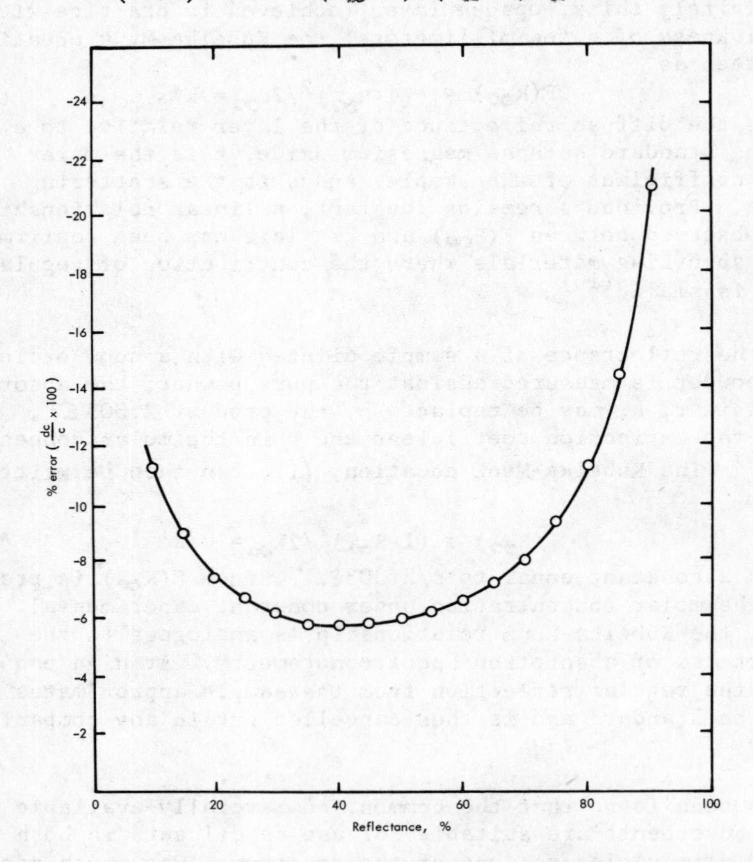

Figure 1.—Percentage error, computed with the use of equation (5), as a function of percentage reflectance.

To determine the value of R_∞ which will minimize the relative error in c, d(% error in c)/dR_∞ is equated to zero. The positive solution of the resulting equation,

$$R_\infty^2 + 2R_\infty - 1 = 0, \qquad (6)$$

indicate that the minimum relative error in c occurs at a reflectance value of 0.414, corresponding to a reflectance reading of 41.4%R. This is presented graphically in Fig. 1, in which the % error, computed by equation (5), is plotted as a function of the % reflectance. As is shown there, the minimum in the resulting curve corresponds to the 41.4% R value obtained by the solution of equation (6).

Reflectance spectrophotometric methods of analysis usually involves the use of a calibration curve prepared by measuring the reflectance of samples containing known amounts of the substance of interest. Conformity to the Kubelka-Munk equation is indicated if the data, when plotted in the form $F(R_\infty)$ vs. c, take the form of a straight line. From the standpoint of utility in analysis, however, it is immaterial whether the system in question conforms

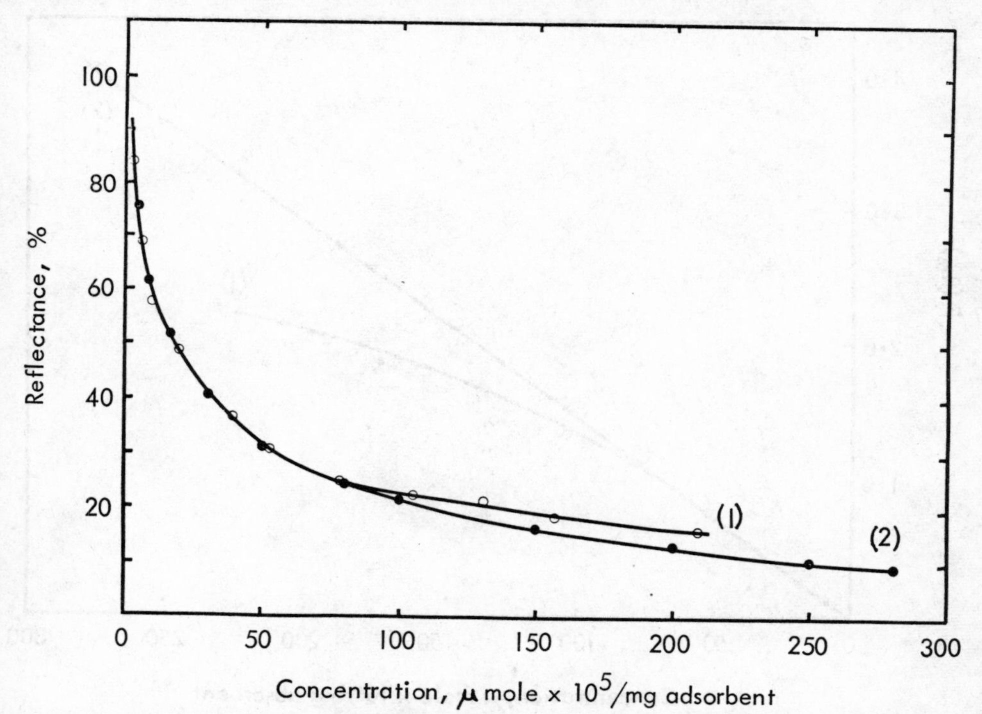

Figure 2.--Percentage reflectance at 545 mμ of Rhodamine B adsorbed on silica gel, as a function of concentration.
o—Experimental values.
●—Theoretical values.

to the Kubelka-Munk equation or not. Provided some sort of near-linear relationship is found to exist between reflectance and concentration, it is more important, if the method is to be used for analysis, to select a suitable concentration range for the analysis and to evaluate its accuracy.

To illustrate how this might be done, Rhodamine B adsorbed on silica gel G can be used as a model system. If the data obtained for this system,[13] which are presented as curve (1) of Fig. 2, are plotted in the form $F(R_\infty)$ vs. c, curve (1) of Fig. 3 results. Although the latter is linear over a considerable portion of the concentration range investigated, it does not conform to the Kubelka-Munk equation at higher concentrations. This becomes more evident when the two experimental curves are contrasted with the pair of hypothetical curves that would have been obtained had the system behaved ideally--curves (2) of Figs. 2 and 3. Such a departure from linearity has been ascribed to the approaching saturation of the adsorbent surface by the first monomelecular layer of the adsorbed species.[11]

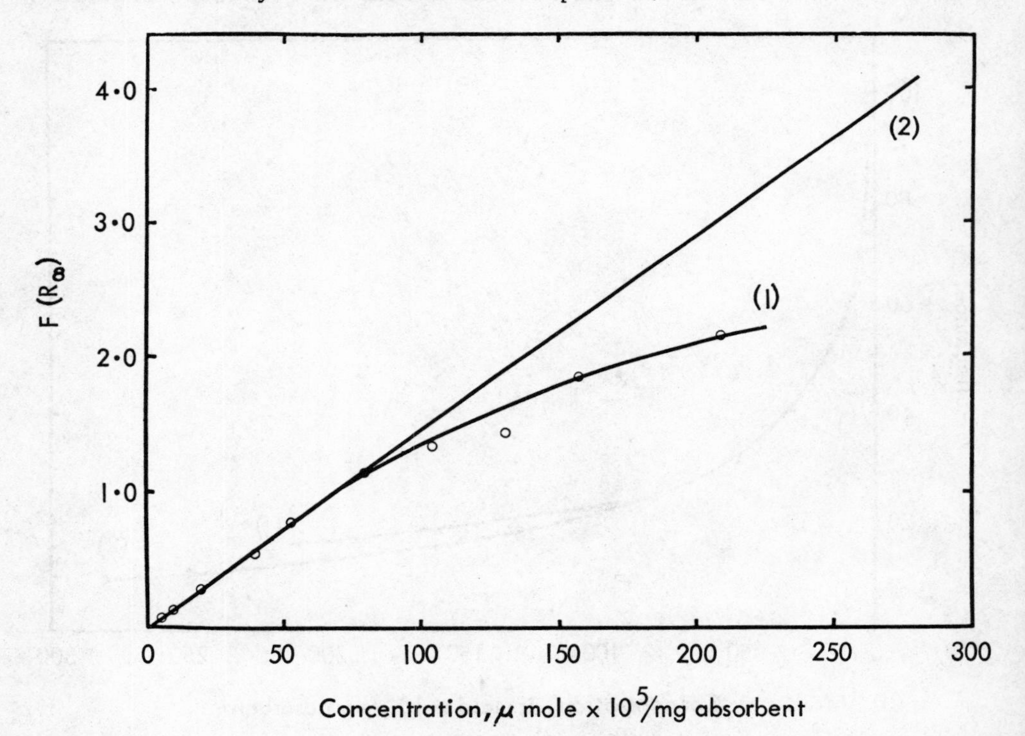

Figure 3.—Kubelka-Munk values at 545 mμ of Rhodamine B adsorbed on silica gel, as a function of concentration.

o —Experimental values.

● —Theoretical values.

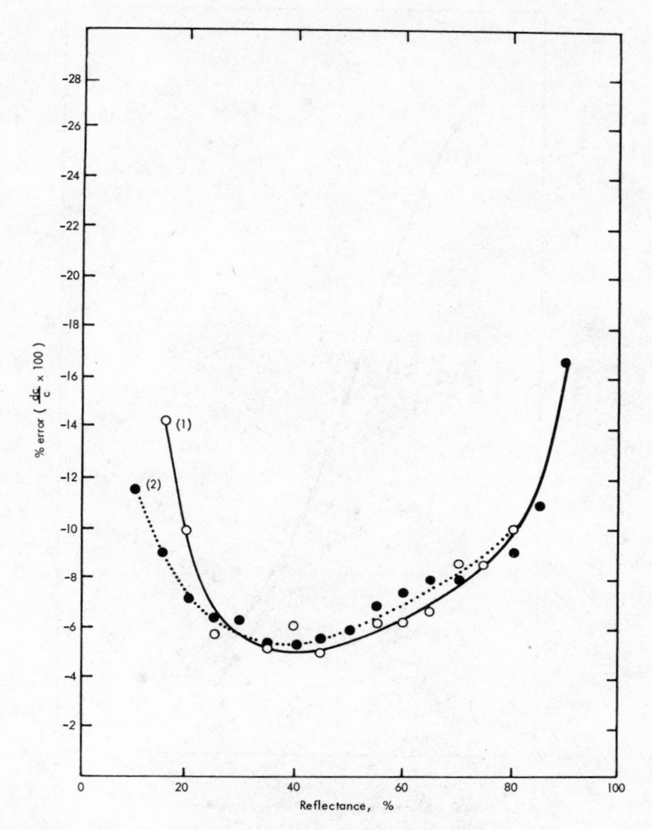

Figure 4.—Percent error arising from a reading error of 1%R (estimated graphically) as a function of percent reflectance.
o—Experimental Rhodamine B values.
●—Hypothetical Rhodamine B values.

When the relative error in concentration arising from a 1% error in R is plotted against % reflectance, as in Fig. 4, there is close agreement between the theoretical curve of Fig. 1 and the curve for hypothetical values for the Rhodamine B system. The scatter in curve (2) arises from the method of calculation. There is poorer agreement with the experimental curve because of the departure from the Kubelka-Munk equation. Nevertheless, the curves do show the range of reflectance that gives minimum error.

Another approach to the selection of the optimum range for reflectance analysis is one suggested by Ringbom,[14] and later Ayres,[15] who evaluated relative error and defined the suitable range for absorption analysis by plotting the absorptancy (1-transmittance) against the logarithm of the concentration. When the reflectance data obtained with Rhodamine B and the hypothetical

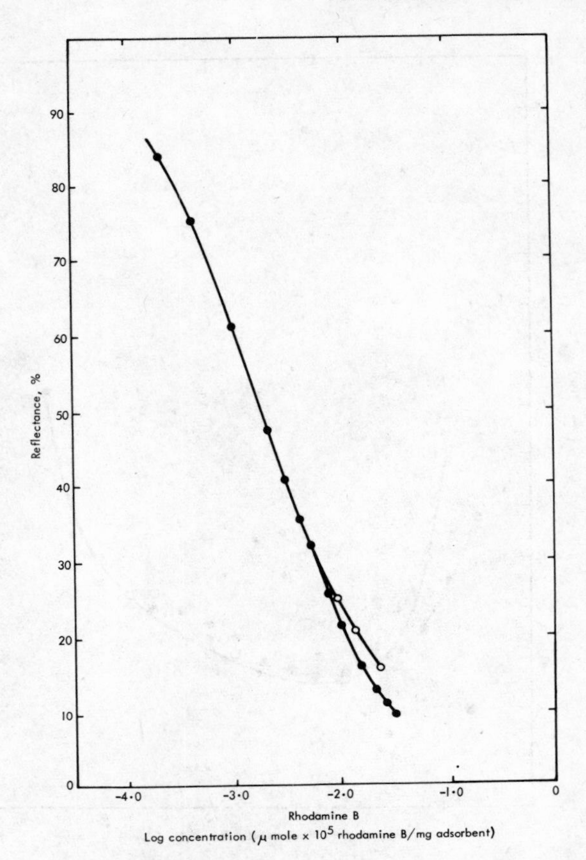

Figure 5.—Percent reflectance as a function of the logarithm of concentration.

o—Experimental Rhodamine B values.
●—Hypothetical Rhodamine B values.

data are plotted as % reflectance vs. the logarithm of concentration (Fig. 5), the advantages of this method become manifest. The optimum range for analysis corresponds to that portion of each curve exhibiting the greatest slope. Since a considerable portion of each of the curves shown in Fig. 5 is also reasonably linear about this region, it is apparent that good accuracy can be expected over a wide concentration range. The maximum accuracy can be estimated from the equation

$$(dc/c)\text{x}100 = 2.303\ d(\log c)\ \text{x}\ 100 = 2.303\left[d(\log c)/dR\right]\text{x}100dR. \quad (7)$$

Assuming a constant reading error of 1%R, i.e., dR = 0.01,

$$\%\ \text{error} = 2.303\ \left[d(\log c)/dR\right]. \quad (8)$$

For the systems under consideration the optimum range for analysis can be arrived at by a consideration of the curves in Fig. 5, and the % error resulting from a reading error of 1% R can be computed with the use of equation (8) and the slope of the appropriate

Table I. Optimum Range for Analysis and % Error

Curve	System	Optimum range, (%R)	% Analysis error for 1% reading error
1	Experimental Rhodamine B	25-65	6.0
2	Hypothetical Rhodamine B	20-65	6.0

curve. When this is done, the data obtained, which are presented in Table I, are found to be in accord with those resulting from the application of the graphical method described first. Values for the hypothetical system also agree with the figures obtained for an ideal system and computed with the use of equation (5).

It would seem, therefore, that the minimum error in reflectance spectrophotometric analysis is about 6% per 1% R reading error, regardless of whether the system conforms to the Kubelka-Munk equation or not. This value can obviously be decreased by reducing the reading error. Although a reading error amounting to 0.5%R should not be too difficult to attain, it would be unrealistic to expect a precision better than 0.1-0.2%R, and, therefore, a minimum error smaller than 1-2%.

Similarly, regardless of whether a system conforms to the Kubelka-Munk equation, the optimum range for analysis can be arrived at after plotting the reflectance data according to either of the two procedures described. The Ringbom method has the advantage of not only making available the optimum range and maximum accuracy, but also of providing a plot usable as a calibration curve.

APPARATUS

A Beckman Model DK-2 Spectrophotometer fitted with a standard reflectance attachment was used to locate ultraviolet-absorbing compounds resolved on chromatoplates and to record reflectance spectra of resolved compounds. All other reflectance measurements were made with a similarly outfitted Beckman DU Spectrophotometer. Three types of reflectance cells, each capable of accommodating samples weighing as much as 100 mg, have been devised to carry reflectance measurements on spots following their removal from thin-layer plates.

(1) Glass Window Cell.[12] This cell consisted of white paperboard to which a 3.7 x 2.5 x 0.1 cm microscope cover glass had been affixed with two pieces of masking tape. The white backing paper was cut to a size (4.0 x 3.0 x 0.1 cm) which permitted its introduction into the sample holder of the reflectance attachment of the Beckman DU Spectrophotometer. These data are presented

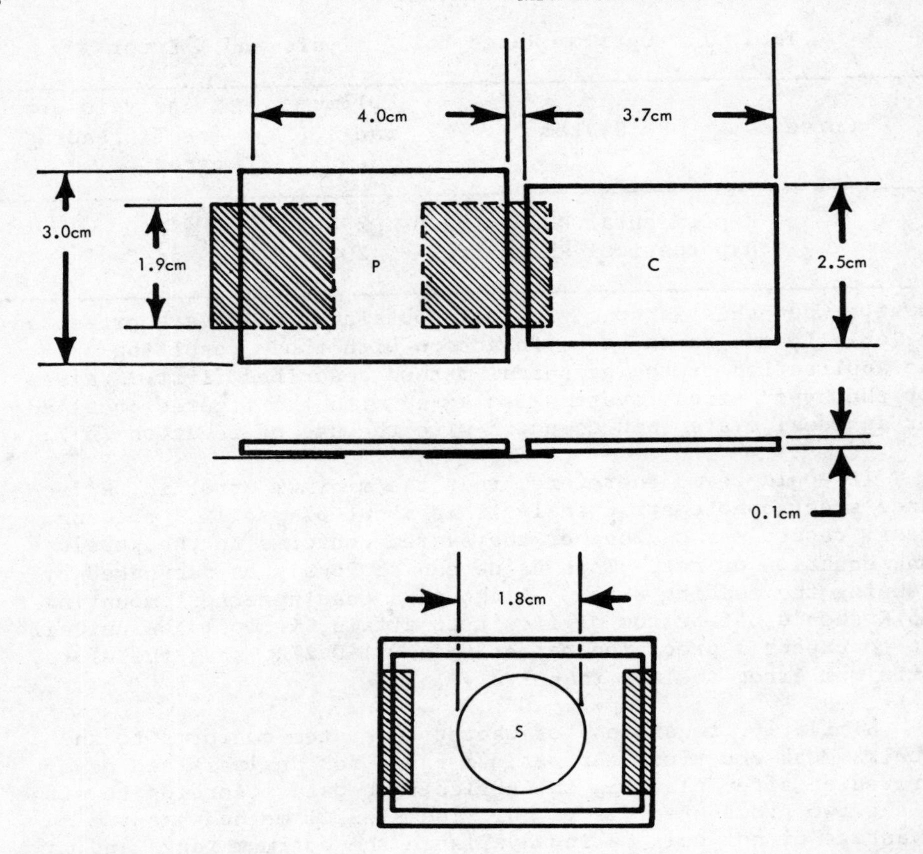

Figure 6. Glass window cell. Dimensions of cell elements and sketch of assembled cell. P, backing paper; c, microscope cover glass; s, sample.

schematically in Fig. 6, as is a sketch of the assembled cell. The analytical sample consisting of adsorbent plus the resolved compound was carefully compressed between the cover glass and the paperboard until a thin layer was obtained.

(2) Quartz window cell.[16] This cell consisted of a circular quartz plate, which had a diameter of 22 mm, superimposed on a 40 x 40 x 1 mm piece of white paperboard. The quartz disk was held in place by means of a 40 x 40 x 3 mm plastic plate which was affixed to the backing paper with two pieces of masking tape. A circular window, 19 mm in diameter, in the upper surface of the plate opened into a concentric circular well, 24 mm in diameter, that was deep enough to accommodate the quartz disk. These data are presented schematically in Fig. 7, where a sketch of the assembled cell is also given. The analytical sample, after being introduced into the cell, was carefully compressed between the

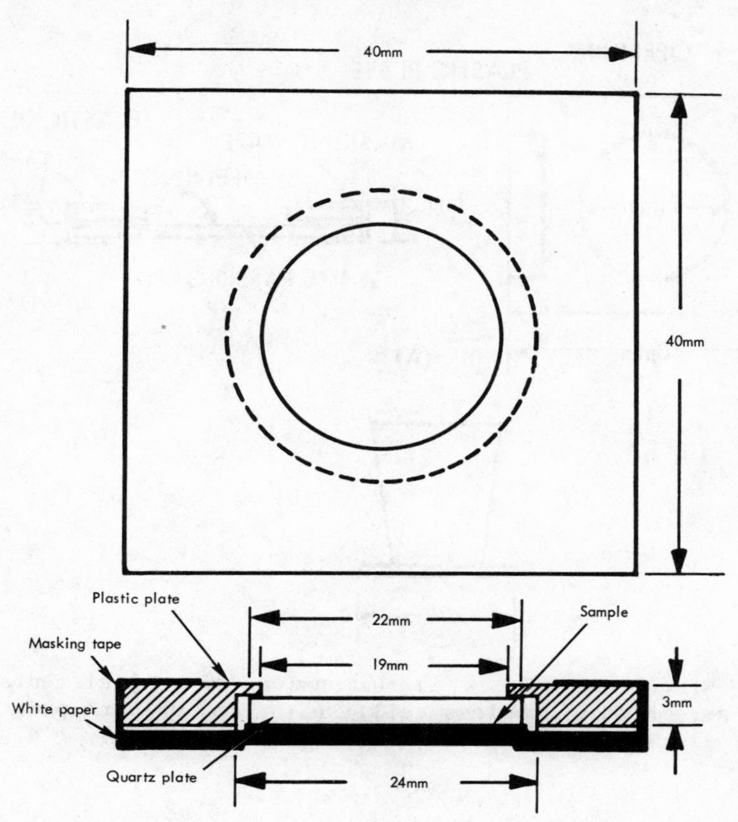

Figure 7. Quartz window cell. Dimensions of cell elements and sketch of assembled cell.

quartz disk and the paperboard by rotating the former until a thin layer was obtained.

(3) Windowless cell.[17] This cell consisted of a 35 x 40 mm plastic plate affixed to white paperboard of similar dimensions with two pieces of masking tape. As shown in Fig. 8, which presents the dimensions of the cell elements as well as a sketch of the assembled cell, the plastic plate had a circular opening, 21 mm in diameter, in its center. The cell was packed by introducing the sample into the opening and then compressing it with a fitted tamp made of an aluminum planchet affixed to a cork stopper.

Obviously the glass window cell can be employed only with samples that absorb in the visible while the other two are suitable for use in both the visible and ultraviolet regions of the spectrum. Of the latter two, the windowless cell, though it is simpler in design and costs almost nothing to make, provides less protection to the analytical sample.

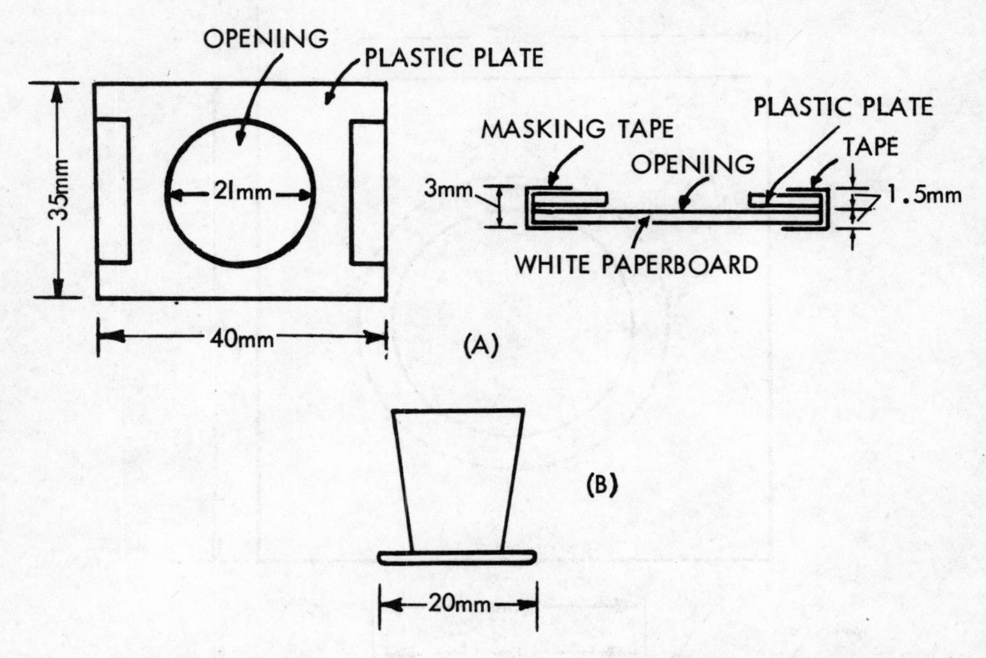

Figure 8. Windowless cell. (A) Dimensions of cell elements and sketch of assembled windowless cell. (B) Tamp used to pack cell.

GENERAL PROCEDURE

Chromatographic Separations

The application of spectral reflectance to thin-layer chromatography imposes no limitation on the type of chromatographic separation procedure that can be used. A variety of procedures for the resolution of a number of mixtures on different commercially available thin-layer adsorbents has been employed in conjunction with spectral reflectance analysis with little or no modification. In these investigations, solutions of substances undergoing analysis were applied as spots by means of graduated micropipets to 20 x 5 x 0.35-cm (for one-dimensional analysis) and 20 x 20 x 0.35-cm (for one- or two-dimensional analysis) plates. A Desaga-Brinkman Model SII applicator was used to coat the plates which were allowed to "set" at room temperature, dried, and then stored in a desiccator before use. The plates were dried after spotting and then developed by the ascending technique with the use of an appropriate solvent system. The developed plates were then redried and readied for analysis.

Location of Resolved Compounds

Unlike colored compounds whose position on the plate following resolution is readily apparent, colorless compounds must be located by resorting to one of several techniques developed for this purpose. Usually the plate is treated with a chromogenic reagent. Somewhat less often a luminous pigment is added to the adsorbent, or, if the compound in question fluoresces, the plate is observed under ultraviolet illumination. Occasionally it is possible to form a colored product by heating the plate.[16] Compounds absorbing in the ultraviolet have been located by scanning the thin-plates with a Beckman Model DK-2 Spectrophotometer set at the absorption maximum of the compound of interest.[17] The scanning was carried out by holding the chromatoplate, which was taped to a protective plastic plate, against the sample exit port of the reflectance attachment unit in such a way that the adsorbent along the path of chromatographic development was exposed to the impinging beam of light. As may be seen in Fig. 9, the 0.3-cm-thick

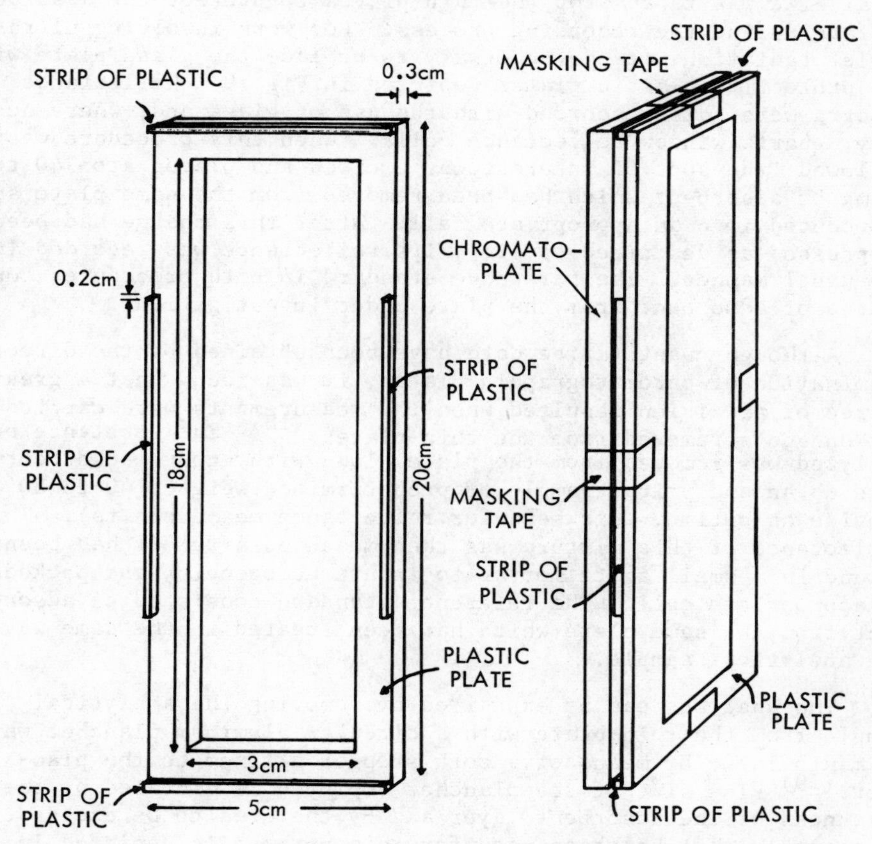

Figure 9. Assembly employed for scanning chromatoplates.

plastic plate, whose other dimensions matched those of the chromatoplates, had a 3 x 18-cm window about which were spaced four 0.1 x 0.2-cm strips of plastic. A sudden decrease in reflectance occurred when the beam of light fell upon a spot containing the compound of interest. During the scanning process, the reflectance attachment was covered with a dark cloth to exclude outside light. When a particular compound had been located, its position was marked on the reverse side of the glass plate.

Recording of Reflectance Spectra and the Measurement of Reflectance.

Once the resolved compounds had been located, their reflectance spectra were recorded in one of two ways. The direct recording of spectra was made possible by positioning chromatoplates against the sample exit port of the reflectance attachment of a Beckman Model DK-2 Spectrophotometer in such a way that the light beam was centered on the spot of interest. A glass plate of identical size was taped atop the thin-plates to protect the adsorbent surface during the recording process. For work involving ultraviolet radiation, it was necessary to replace the glass plate with the protective plastic plates depicted in Fig. 9. Reflectance spectra were also recorded with the use of glass and, where necessary, quartz window reflectance cells. When this procedure was followed, the spot of interest was excised and placed atop 40 to 50 mg of adsorbent which had been removed from the same plate and introduced into an appropriate cell. After this charge had been compressed as described earlier, its reflectance was recorded in the usual manner. The reference standard in both procedures consisted of adsorbent from the plate under investigation.

Although quantitative data have been obtained by the direct examination of chromatographic plates, it was found that a greater degree of precision resulted when the measurements were carried out on spots removed from the thin-plates.[18] The substance being analyzed was removed from the plate along with enough adsorbent to make up an analytical sample of predetermined weight (40 to 80 mg provide an optimum thickness for reflectance measurements). The reflectance of this mixture was then measured after it had been ground in a small agate mortar to insure homogeneity and packed in an appropriate cell. The reference standard consisted of adsorbent from the same plate which had been treated in the same way as the analytical sample.

The analysis can be expedited by removing the analytical sample from the thin-plate with a circular aluminum planchet which is manipulated by means of a cork stopper affixed to the planchet.[19] The size of the planchet employed is dictated by the thickness of the adsorbent layer and by the area to be excised. An assembly that has been used for this purpose is depicted in Fig. 10. Once the sample had been cut out of an adsorbent layer

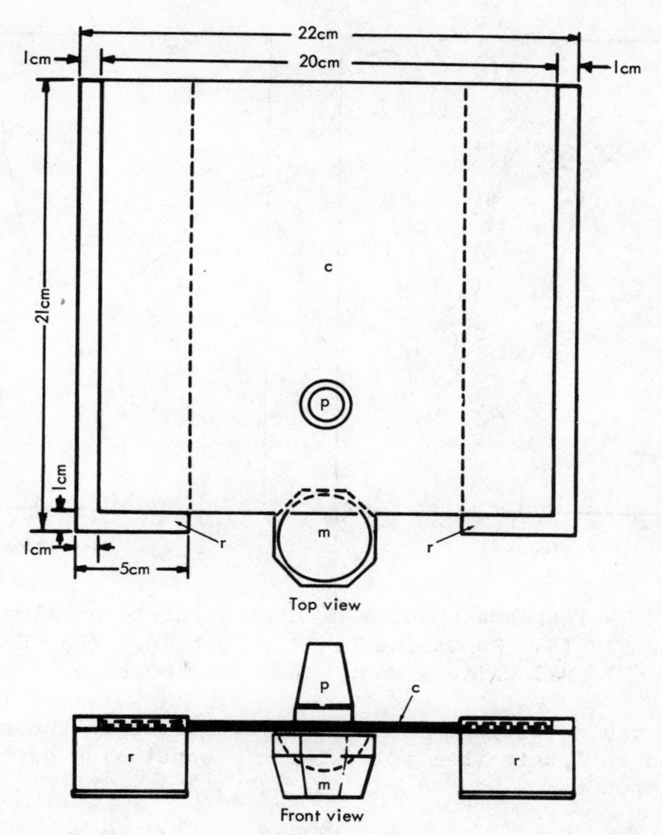

Figure 10. Assembly employed to excise samples from chromatoplates. C, chromatoplate; m, agate mortar; p, planchet affixed to cork stopper; r, wooden rack.

by exerting a slight pressure on the inverted planchet, the most direct path between it and the nearest plate edge was cleared of adsorbent with a brush and the planchet was moved along this path until the sample was deposited in the agate mortar. By means of this procedure, it is possible to remove a spot for analysis in approximately 30 sec. It has also been shown that the surfaces laid down by a commercial applicator are with certain limitations uniform enough to permit this procedure to be employed in routine analyses without sacrificing the precision inherent in the reflectance technique.

APPLICATIONS

Spectral reflectance has been used for the in situ identification and determination of a variety of substances following their resolution on thin-plates. For purposes of discussion, these

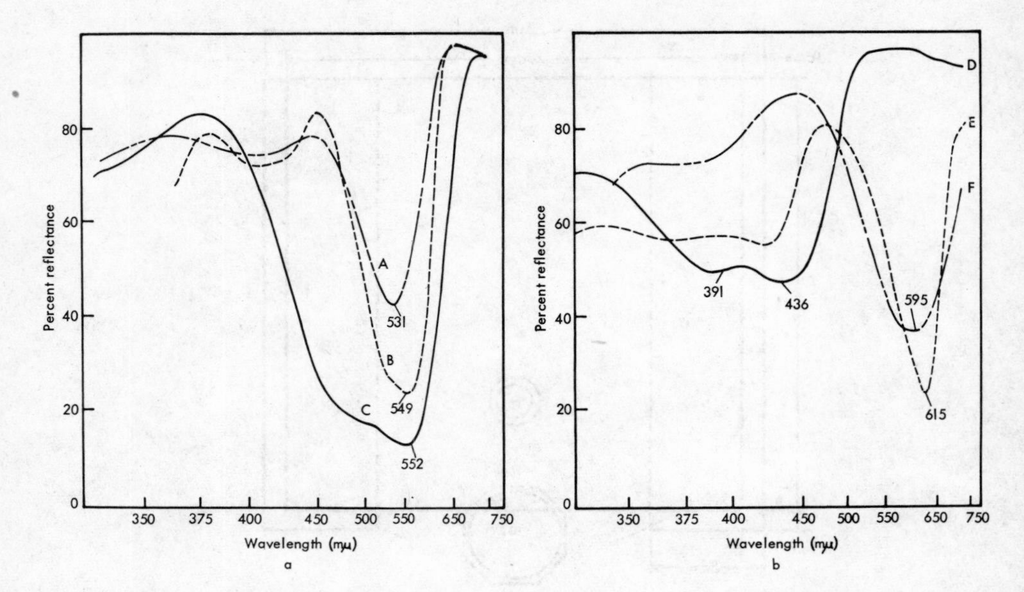

Figure 11. Reflectance spectra of dyes adsorbed on alumina. (A) Eosine B. (B) Rhodamine B. (C) Fuchsin. (D) Naphthol yellow S. (E) Malachite green. (F) Aniline blue.

substances can be divided into three categories on the basis of the method used to locate them following the conclusion of the chromatographic process.

(1) *Colored substances*. The analysis of colored compounds, particularly if they are stable, usually presents no difficulty. For example, not only was it possible by the direct examination of chromatoplates to identify the components of mixtures of dyes separated on the plates rapidly and without recourse to R_f values, but the same operation was also capable of providing quantitative data having a precision of approximately ±5%.[18] Fig. 11 depicts the reflectance spectra which were obtained for six water-soluble dyes adsorbed on alumina. A precision of about 3% was attained if the reflectance measurements were carried out on spots removed from chromatographic plates and packed in an appropriate cell. When this procedure was employed as the basis for a student experiment in an undergraduate course in quantitative analysis, the data obtained were such as to indicate that the reflectance technique can be employed successfully even by individuals having no specific skills or prior experience.[20]

(2) *Colorless substances which react with a suitable chromogenic reagent to generate a color*. Typical of the approach that can be employed with this category of compounds is a procedure devised

for the determination of amino acids resolved on chromatoplates which depends on the use of a ninhydrin spray reagent described by Bull et al.[21] The method was later improved by adding the chromogenic reagent to the developing solvent.[22] In addition to facilitating the determination, the elimination of the spraying operation with its attendent irregularities resulted in a substantial increase in precision and accuracy. Relative errors of approximately 4% and sensitivities between 1×10^{-8} and 5×10^{-9} moles were reported for most of the amino acids investigated. When a modified ninhydrin spray reagent was employed to generate various colors, it was possible to identify eighteen amino acids resolved on chromatoplates by means of reflectance spectroscopy used in conjunction with R_f values and visual observation.[23] This procedure enabled one- and two-dimensional chromatograms to be read rapidly and accurately without requiring the conditions necessary for the attainment of reproducible R_f values.

The feasibility of this approach was also demonstrated by the successful determination of mixtures of sugars using aniline and trichloroacetic acid as a spray reagent.[24]

(3) Colorless substances which have characteristic ultraviolet spectra. The first application of ultraviolet reflectance spectroscopy to thin-layer chromatography involved the determination of the composition of mixtures of aspirin (acetyl salicylic acid) and salicylic acid which had been separated on silica gel plates.[16] In that instance, however, the location of the resolved components depended on the formation of colored substances resulting from the heating of the developed thin-plates.

The technique was subsequently applied to the nondestructive analysis of mixtures of some vitamins of the B group[17] and of nucleotides.[25] Because these compounds are colorless, it was first necessary to locate them once they had been resolved on thin-plates. Two methods were used to locate the resolved vitamins. One involved the addition of a luminous pigment to the silica gel G adsorbent and examination of the chromatoplates under ultraviolet illumination. The other employed direct scanning of silica gel G chromatoplates by means of a spectrophotometer fitted with a standard reflectance attachment in the manner described earlier. The location of the resolved nucleotides, which fluoresce, was achieved by examination of the chromatoplates under ultraviolet illumination. In both investigations, all but a few of the compounds could be identified unequivocally by means of their reflectance spectra, with the ones having similar spectra being distinguished with the aid of their R_f values. Table II, which lists the absorption maxima and R_f values of some vitamins adsorbed on silica gel G and on silica gel G-luminous pigment mixture, indicates how this can be done in the case of the vitamins. It is noteworthy that the spectra and R_f values obtained with the two

Table II. Absorption Maxima and R_f Values of Vitamins Adsorbed on Silica Gel G and on Silica Gel G-Luminous Pigment Mixture

Vitamin	Absorption maximum (mμ)		R_f value	
	Silica gel G-lum. pig. mixture	Silica gel G	Silica gel G-lum. pig. mixture	Silica gel G
Thiamine hydrochloride	278	278	0.00	0.00
Pyridoxine hydrochloride	298	298	0.17	0.18
Nicotinamide	263-268	264	0.51	0.51
Nicotinic acid	262-268	264	0.66	0.65
p-Aminobenzoic acid	295	295	0.85	0.86

adsorbent systems are in such close agreement that it should be possible to use them interchangeably for most purposes. Except for the type of cell employed, the procedure followed in gathering quantitative data was the same as that used with systems absorbing in the visible portion of the spectrum.

Except for manifestly unstable systems, close agreement was obtained among the calibration plots prepared from dilution series data gathered at different times. This would seem to indicate the feasibility of carrying out routine analyses by means of this technique without preparing a new calibration curve for each set of determinations.

Although the studies carried out thus far have been restricted to only a few categories of compounds, there appears to be no reason why the technique cannot be applied directly or with some slight modification to other organic, biological or even inorganic systems.[26-28] The application of the method to the analysis of trace metals in sea water is currently in progress.

REFERENCES

1. E. H. Winslow and H. A. Liebhafsky, Anal. Chem., 21:1338 (1949).

2. K. Yamaguchi, S. Fujii, T. Tabata and S. Kato, J. Pharm. Soc. Japan, 74:1322 (1954).

3. K. Yamaguchi, S. Fukushima and M. Ito, J. Pharm. Soc. Japan, 75:556 (1955).

4. R. B. Fischer and F. Vratny, Anal. Chim. Acta, 13:588 (1955).

5. F. Pruckner, M. von der Schulenburg, and G. Schwuttke, Naturwiss., 38:45 (1951).

6. G. Schwuttke, Z. Angew. Phys., 5:303 (1953).

7. C. A. Lermond and L. B. Rogers, Anal. Chem., 27:340 (1955).

8. P. Kubelka and F. Munk, Z. Techn. Physik, 12:593 (1931).

9. P. Kubelka, J. Opt. Soc. Am., 38:448, 1067 (1948).

10. G. Kortüm and H. Schöttler, Z. Electrochem., 57:353 (1953).

11. G. Kortüm, Angew. Chem. Intern. Ed. Eng., 2:333 (1963).

12. R. W. Frei and M. M. Frodyma, Anal. Chim. Acta., 32:501 (1965).

13. V. T. Lieu and M. M. Frodyma, Talanta, 13:1319 (1966).

14. A. Ringbom, Z. Anal. Chem., 715:332 (1939).

15. G. H. Ayres, Anal. Chem., 21:652 (1949).

16. M. M. Frodyma, V. T. Lieu and R. W. Frei, J. Chromatog., 18:520 (1965).

17. M. M. Frodyma and V. T. Lieu, Anal. Chem., 39:814 (1967).

18. M. M. Frodyma, R. W. Frei and D. J. Williams, J. Chromatog., 13:61 (1964).

19. V. T. Lieu, R. W. Frei, M. M. Frodyma and I. T. Fukui, Anal. Chim. Acta., 33:639 (1965).

20. M. M. Frodyma and R. W. Frei, unpublished data.

21. H. B. Bull, J. W. Hahn and V. R. Baptist, J. Am. Chem. Soc., 71:550 (1949).

22. M. M. Frodyma and R. W. Frei, J. Chromatog., 17:131 (1965).

23. R. W. Frei, I. T. Fukui, V. T. Lieu and M. M. Frodyma, Chimia, 20:23 (1966).

24. F. J. Thaller, M. S. Thesis, Univ. of Hawaii (1965).

25. V. T. Lieu, M. M. Frodyma and L. S. Higashi, Anal. Biochem., 19:454 (1967).

26. R. W. Frei and D. E. Ryan, Anal. Chim. Acta, 37:187 (1967).

27. D. F. Zaye, R. W. Frei and M. M. Frodyma, Anal. Chim. Acta, 39:13 (1967).

28. M. M. Frodyma, D. F. Zaye and V. T. Lieu, Anal. Chim. Acta, (in Press).

REFLECTANCE VARIABLES OF COMPACTED POWDERS

*Elihu A. Schatz

American Machine & Foundry Company

Alexandria, Virginia

Abstract

Numerous factors were found to modify the spectral reflectance of compacted powders in the 0.23 to 2.65μ range. Particular variables of importance were opacity of the particles, degree of compaction, chemical composition and temperature. Other variables studied included particle size, liquid impurities, and effect of vacuum. Transparent particles, such as white oxides, underwent a large decrease in reflectance with increasing compaction, while opaque particles, such as metals and dark nonoxides, underwent a small increase. Particle size variations from 30 to 105μ did not change the reflectance of compacted pure materials. However, particle size was of importance when considering mixtures, since it controlled the relative surface areas of the components. Equations were developed to enable approximate calculation of the reflectance of compacted powder mixtures. Liquid additions to the powders were found to decrease the reflectance of the compacts, and the decrease was greatest for materials having transparent particles. Heating of compacted oxides resulted in either a decrease or increase in reflectance depending on the material and the temperature. Exposure of specimens to vacuum usually resulted in random small variations in reflectance.

1.0 INTRODUCTION

In performing systematic reflectance studies, use of compacted powders allows careful control of numerous variables.

---

*present address: Melpar, Inc., Falls Church, Virginia

For example, particle size, separation between particles, composition, and crystal structure are more easily controlled than with sintered specimens or coatings. By studying the reflectance of compacted powders we have been able to clarify the effects on reflectance of many of the above factors. Principles thus derived can then be used to explain more complex systems, such as sintered specimens and coatings, where chemical and physical interaction of the particles readily occurs.

2.0 EXPERIMENTATION

Spectral reflectance measurements in the 0.23 to 2.65μ range were performed on a Beckman DK-2A Spectroreflectometer. Smoked MgO standards, having a reflectance of about 0.95 over nearly the entire wavelength range, were prepared daily. The reflectance curves are presented relative to these standards. Specimen holders for compacted powders were prepared by machining cavities into stainless steel disks. Most work was performed using a 7/8" diameter cavity about 1/16" deep. Pressures used in sample preparation ranged from about 1000 to 70,500 psi and were applied with highly polished rams. The powders were usually better than 99% pure, and were obtained commercially.

3.0 PRESSURE EFFECT

Initial studies revealed that the preparation pressure had a very significant effect on the reflectance of the compacted powders, and the effect was dependent on the transparency of the particles. Materials, such as white oxides, having relatively transparent particles, exhibited a large decrease in reflectance with increasing pressure. For example, compacted Y_2O_3 underwent a decrease in reflectance of 25 percent at 0.28μ as shown in Fig. 1. Other oxides for which this same trend was obtained included Al_2O_3, $BaSO_4$, CeO_2, MgO, NiO, Sm_2O_3 and ZnO.

The decrease in reflectance with increasing pressure can be explained as follows. Typical particles used in preparing compacted powders are multisided crystals, of which the sides have random orientation. Air, with a refractive index much less than the powder, is found between the compacted particles. When a beam of light impinges on the surface of the compacted powder, some of the light that enters the crystals is incident internally on the particle surfaces at angles greater than the critical angle. Usually this would result in total reflection of those rays. However, because of the closeness of neighboring particles, frustrated total reflection[1,2] occurs. This results in the escape from the crystals and transmission by the layer of some of the light that otherwise would have been totally internally reflected. With increasing pressure the distance between the particles de-

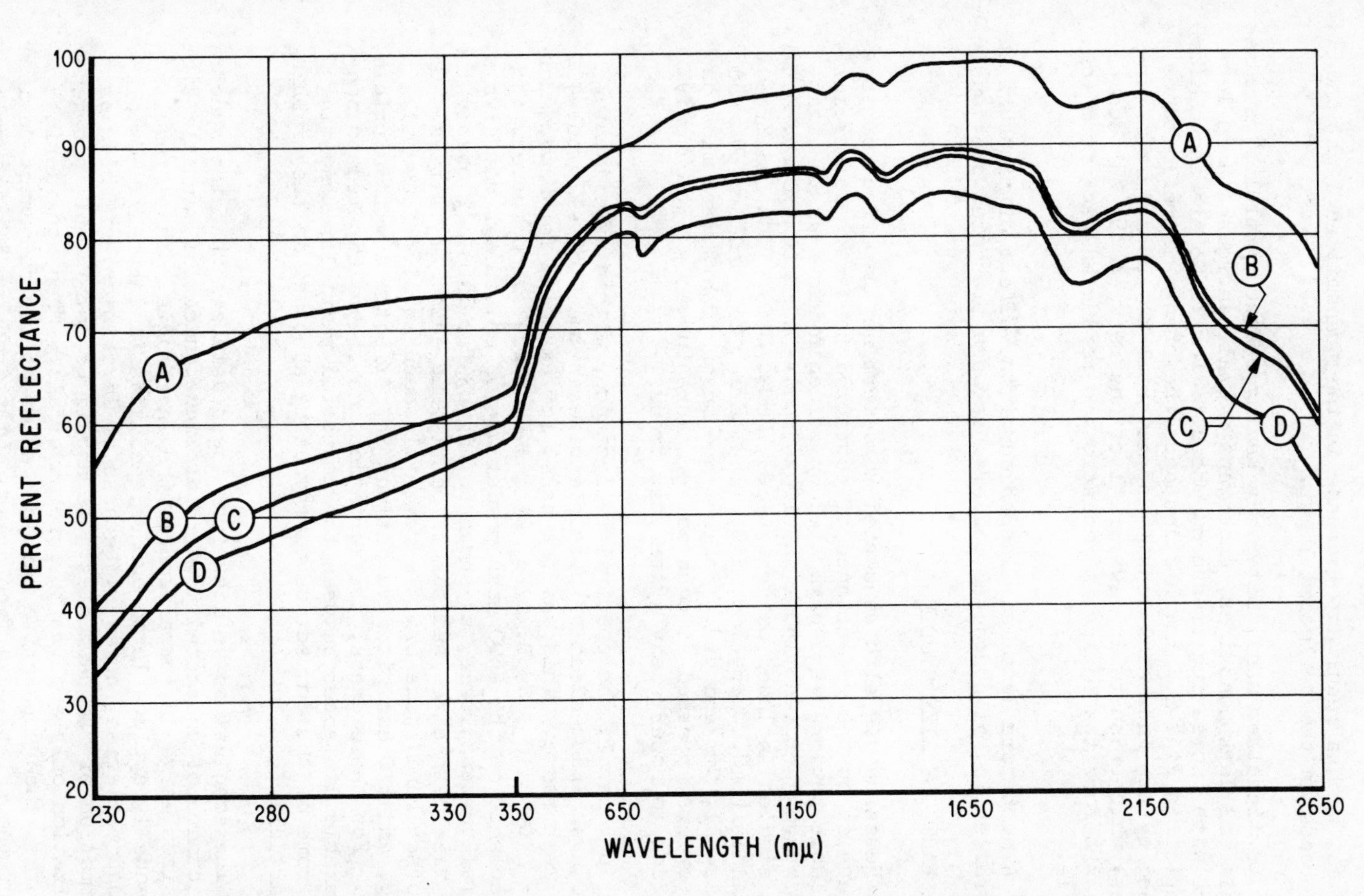

Figure 1. Spectral Total Reflectance (vs. MgO) of Compacted Y_2O_3 Powder: Compaction Pressure (psi), A - 23500; B - 47000; C - 58800; D - 70500.

creases, with a resultant increase in the transmitted light and a decrease in the reflected light.

On the other hand, materials having opaque particles, such as metals and dark nonoxides, underwent a small increase in reflectance with increasing pressure. Results for $TiSi_2$ are shown, for example, in Fig. 2. In this instance the change in reflectance is caused by the increased density of material at the specimen surfaces, which results in fewer reflections between the surface particles before the impinging radiation is reflected back from the sample.

A much more detailed discussion of the effect of pressure on the reflection of compacted powders has been presented previously.(3)

4.0 PARTICLE SIZE STUDIES

Research was also conducted to determine the effect of particle size on the reflectance of compacted powders. Commercially available powders were separated by screening into up to eight size ranges between +150 mesh (105μ) and -500 mesh (30μ). Materials studied included oxides (Al_2O_3, MoO_3, SiO_2, TiO_2 and ZrO_2), nonoxides (B_4C, Cr_2B, MoB, SiC, TiB_2, ZrB_2, ZrC and $ZrSi_2$), and metals (Al, Cr, and Ni). Reflectance measurements were made on the powders, pressed at constant pressure (usually 35,300 psi), in order to observe any systematic trends.

In general, the spectral reflectance, within experimental error, was independent of particle size in the +150 -170 to -400 +500 mesh range. This was the case, for example, for MoO_3, amorphous SiO_2, TiO_2, Cr_2B, TiB_2, Al and Cr powders. Moreover, the reflectance of the -500 mesh powder was often, though not always, several percent lower, as shown in Fig. 3 for TiO_2. Probably the very fine mesh sizes had significantly increased impurities. On the other hand, some brittle materials such as SiO_2 (quartz), $ZrSi_2$, and SiC exhibited variations in reflectance with particle size. For these cases, the variations may be attributed to noncomparable distances between the compacted particles resulting from some of the compaction energy being absorbed in the crushing of particles.

The conclusions we draw from these data are that the spectral reflectance of compacted powders are independent of particle size for sizes large in comparison to the wavelength. These results are equally applicable to oxides, nonoxides and metals. Complications may arise, however, because of impurities, crushing or deformation of the particles. It should be noted that this generalization only applies to pure materials. When considering mixtures, the comparative particle sizes of the component powders affect the relative surface areas, and thereby modify the relative contributions of the components.

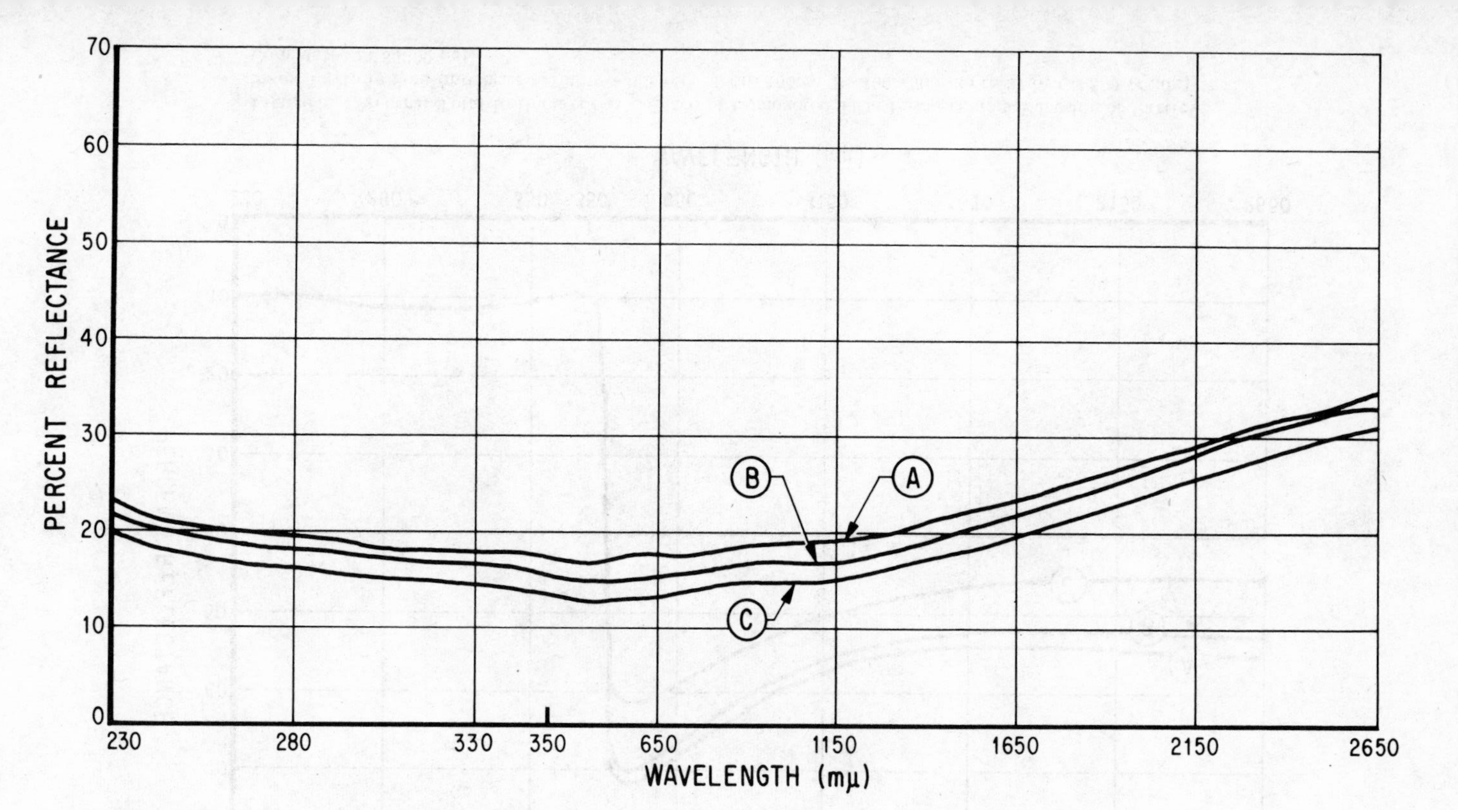

Figure 2. Spectral Total Reflectance (vs. MgO) of Compacted $TiSi_2$ Powder: Compaction Pressure (psi), A - 70500; B - 11800; C - 2350.

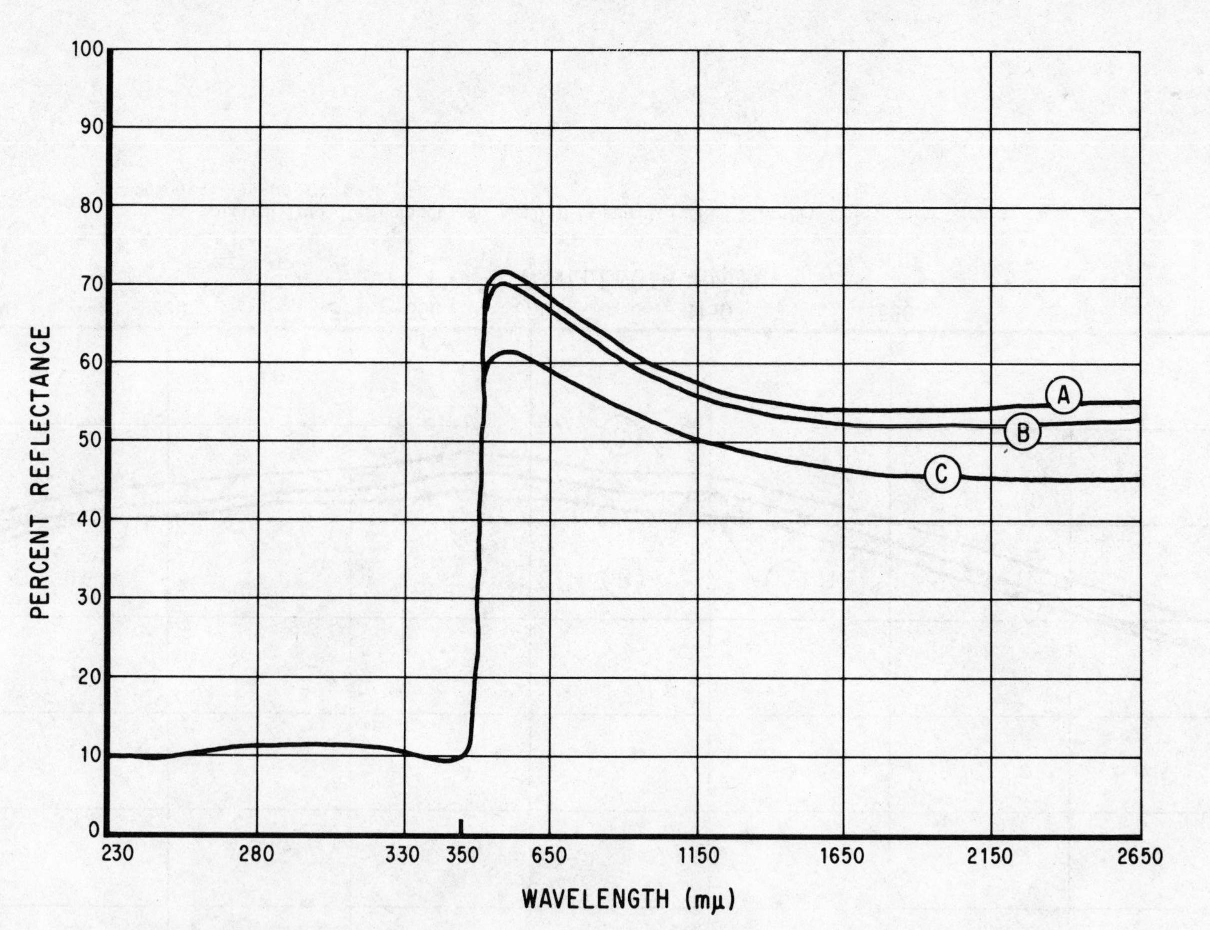

Figure 3. Spectral Total Reflectance (vs. MgO) of Compacted TiO_2 Powders as a Function of Particle Size: Particle Size (mesh), A +150 to -325 +400 (37 to 105μ); B -400 +500 (30 to 37μ); C -500 (< 30μ). Compacted at 35300 psi.

5.0 POWDER MIXTURES

Research was performed to enable the calculation of the spectral reflectance of compacted powder mixtures in terms of the reflectance of the components. In all cases powders of known particle size, usually -230 +270 mesh, were used. The compaction pressure was maintained the same for each mixture studied, usually 23,500 or 35,300 psi, to eliminate possible complications from this variable. The spectral reflectances for varying compositions of over 25 binary mixtures were measured.

Experimental results showed that the mixtures could be classified into three categories of opacity of the powder particles, namely opaque-opaque, opaque-transparent, and transparent-transparent. Metals and dark nonoxides were considered to have opaque particles, while white oxides were considered to have relatively transparent particles. The reflectances of opaque-opaque mixtures, as illustrated for the Al-Ti system in Fig. 4, were always intermediate to those of the components. However, mixtures of opaque-transparent powders could have reflectances either intermediate or considerably lower than for the components. In Fig. 5 the reflectance of a 50 SiO_2-50 Al mixture is much lower than for either pure SiO_2 or Al. On the other hand, transparent-transparent mixtures had reflectances intermediate to those for the components, or at most 3 percent above or below. For example, the reflectance of 10 ZrO_2 -90 Al_2O_3 is higher than the reflectance of the pure components in the 600 to 700 mμ range (Fig. 6).

The reflectance R_M, at any wavelength, for a binary mixture of opaque powders could be expressed in terms of the reflectances R_A and R_B of the components, and their percent surface areas S_A and S_B in the mixture as

$$R_M = S_A R_A + S_B R_B \tag{1}$$

It was necessary to use surface area percent rather than, for example, weight or volume percent, since the reflectance of opaque materials is a surface property. The percent surface area was readily calculated from weight percent composition, given the average density and diameter of the particles. Equation (1) was verified using mixtures of Ni-TiB_2, Ni-B_4C, Al-TiB_2, TiB_2-$ZrSi_2$, Al-Ni and Al-Ti. Agreement within experimental accuracy was obtained in almost all cases. The poorest fit occurred for the Al-Ni mixtures where the calculated values were consistently about 0.06 units greater than the measured values. This exception could be explained as being caused by cold-welding of the Al particles under pressure, resulting in larger particle sizes and smaller surface area for the aluminum than calculated.

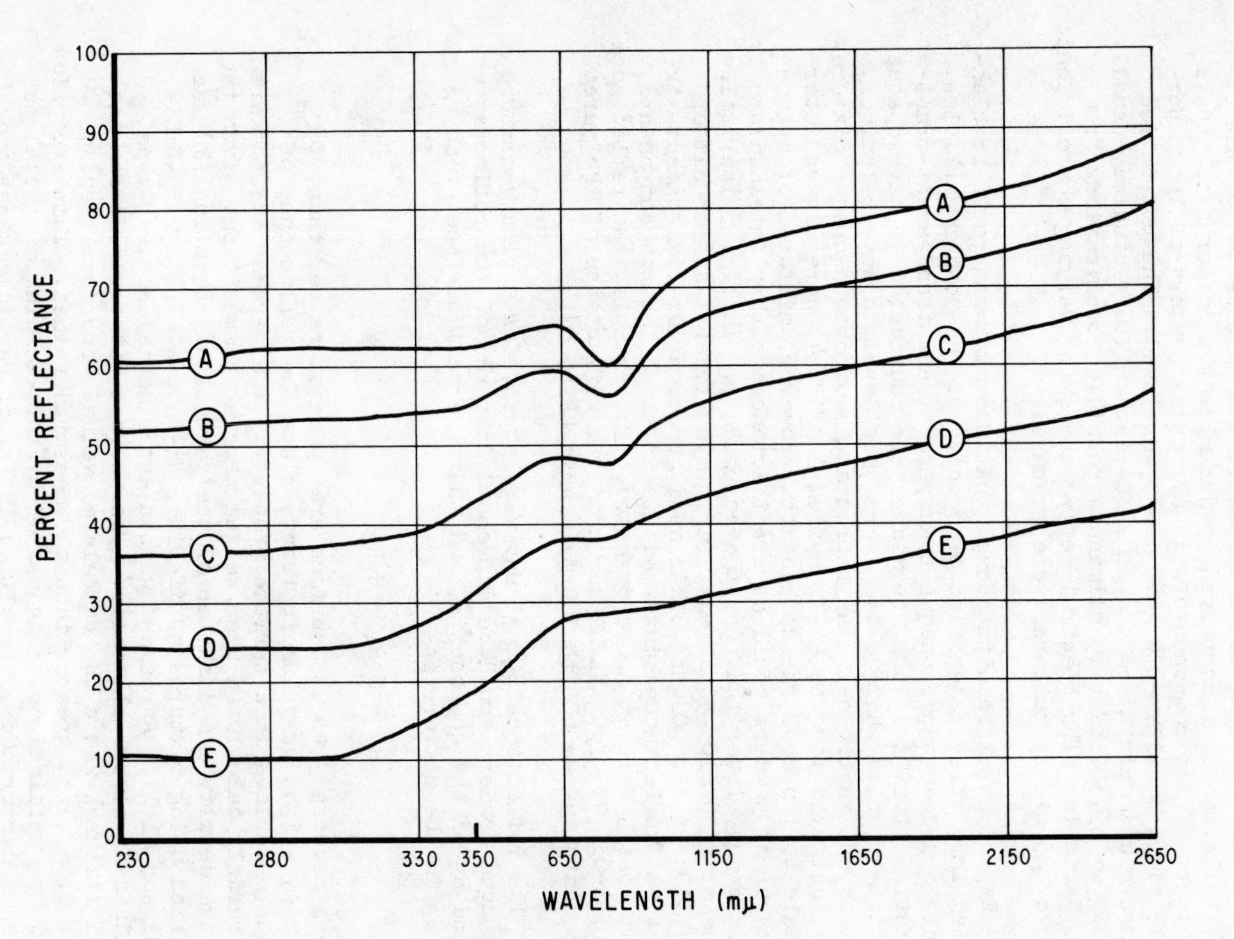

Figure 4. Spectral Total Reflectance (vs. MgO) of Al-Ti Mixtures: Weight Percent, A 100 Al; B 75Al-25Ti; C 50Al-50Ti; D 25Al-75Ti; E 100 Ti. Compacted at 23500 psi, -230 +270 Mesh Powders.

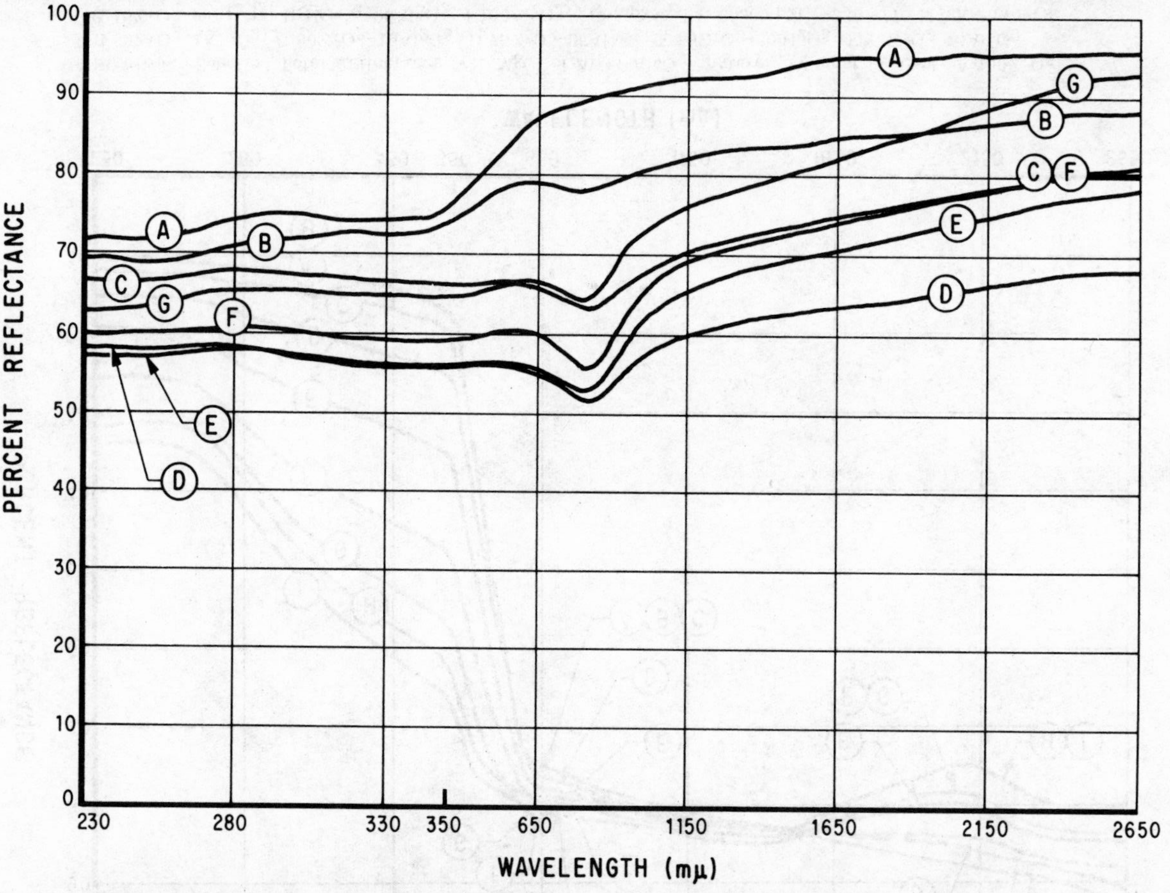

Figure 5. Spectral Total Reflectance (vs. MgO) of SiO_2-Al Mixtures: Weight Percent, A $100SiO_2$; B $90SiO_2$-10Al; C $75SiO_2$-25Al; D $50SiO_2$-50Al; E $25SiO_2$-75Al; F $10SiO_2$-90Al; G 100Al. Compacted at 23500 psi, -230 +270 Mesh Powders.

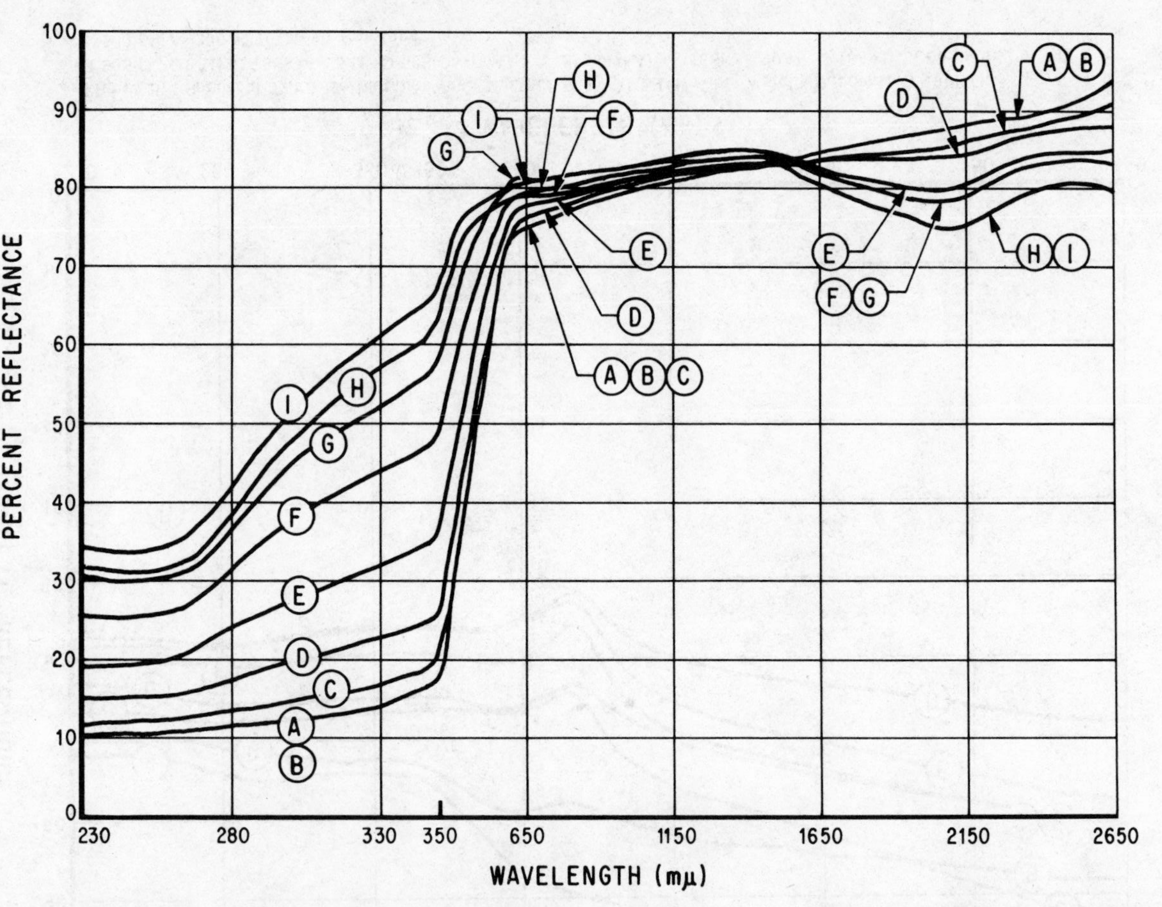

Figure 6. Spectral Total Reflectance (vs. MgO) of ZrO_2-Al_2O_3 Mixtures: Weight Percent, A $100ZrO_2$; B $99ZrO_2$-$1Al_2O_3$; C $90ZrO_2$-$10Al_2O_3$; D $75ZrO_2$-$25Al_2O_3$; E $50ZrO_2$-$50Al_2O_3$; F $25ZrO_2$-$75Al_2O_3$; G $10ZrO_2$-$90Al_2O_3$; H $1ZrO_2$-$99Al_2O_3$; I $100Al_2O_3$. Compacted at 35300 psi, -230 +270 Mesh Powders.

In the case of opaque-transparent mixtures, consideration needed to be given to the fact that the reflectance of transparent particles is dependent both on first surface and internal reflectance. When opaque particles were mixed with transparent powders some transmission of light from one particle to the next was prevented and thereby the internal reflectance contribution of the transparent material was reduced disproportionately. Taking this factor into account leads to the equation

$$R_M = S_A R_A + S_B R_B - S_A S_B (R_A - R_A^S) \quad (2)$$

for the reflectance of an opaque-transparent mixture, where R_A^S, the single-particle reflectance of the transparent material, represents the reflectance of an average particle A when completely surrounded by particles of type B. The approximate value of R_A^S can be calculated from the refractive index n_A according to the Fresnel relationship for normal incidence, $R_A^S=(n_A-1)^2/(n_A+1)^2$. Equation (2) successfully accounts for reflectances lower than for either component. Good agreement with experimental data was obtained for SiO_2-Al, ZrO_2-Al, MoO_3-B_4C, and MoO_3-Ni mixtures, while poorer agreement was obtained for TiO_2-Al, SiO_2-Ni, and SiO_2-TiB_2 mixtures.

The most complex type mixtures, from a theoretical viewpoint, involved transparent-transparent systems, such as oxide-oxide. In developing a suitable equation for such mixtures it was necessary to consider the relative transparencies of the component particles. By assuming, for calculation purposes, that the relative transparencies of the particles were equal to the relative reflectances of the pure components, the following equation was developed:

$$R_M = S_A R_A + S_B R_B + S_A S_B \left[R_A^S (1 - R_B/R_A) + R_B^S (1 - R_A/R_B) \right] \quad (3)$$

Since the third term of the right hand side is small and can be either positive or negative, the equation satisfactorily explains the cases where the reflectances of transparent mixtures are a few percent higher or lower than for the components. Good agreement with experimental data was obtained for Al_2O_3-SiO_2, ZrO_2-Al_2O_3 and MoO_3-SiO_2 mixtures.

In Table I the calculated and measured reflectances for two cases of each type mixture are compared at 650 and 2150 mμ. A more detailed presentation of the data, including a more complete derivation of the above equations, has been presented previously.[4]

TABLE 1.

COMPARISON OF CALCULATED AND EXPERIMENTAL REFLECTANCES FOR SELECTED COMPACTED POWDER MIXTURES

	Mixture	Wt. %	Surface Area %	Reflectance at 650 mμ calc.	Reflectance at 650 mμ meas.	Reflectance at 2150 mμ calc.	Reflectance at 2150 mμ meas.
A. Opaque-opaque	Ni-TiB_2	100 Ni	100 Ni	----	.26	----	.48
		75-25	60-40	.216	.215	.400	.395
		50-50	33-67	.186	.185	.346	.335
		25-75	14-86	.165	.155	.31	.33
		100 TiB_2	100 TiB_2	----	.15	----	.28
	Al-Ti	100 Al	100 Al	----	.65	----	.822
		75-25	83.3-16.7	.588	.594	.748	.743
		50-50	62.5-37.5	.511	.482	.656	.64
		25-75	35.7-64.3	.41	.38	.538	.516
		100 Ti	100 Ti	----	.278	----	.38
B. Opaque-transparent							
	SiO_2-Al	100 SiO_2	100 SiO_2	----	.87	----	.953
	$R^S(SiO_2)$=0.034	90-19	91-9	.783	.79	.87	.876
		75-25	78-22	.68	.66	.78	.78
		50-50	54-46	.568	.55	.69	.66
		25-75	28-72	.554	.555	.72	.75
		10-90	11.5-88.5	.60	.60	.80	.78
		100 Al	100 Al	----	.665	----	.881
	ZrO_2-Al	100 ZrO_2	100 ZrO_2	----	.76	----	.88
	$R^S(ZrO_2)$=0.14	99-1	98-2	.75	.70	.865	.835
		90-10	81-19	.65	.54	.76	.69
		75-25	59-41	.57	.51	.68	.675
		50-50	33-67	.56	.56	.68	.73
		25-75	14-86	.61	.63	.75	.785
		10-90	5-95	.65	.655	.80	.82
		1-99	0.5-99.5	.668	.655	.826	.82
		100 Al	100 Al	----	.67	----	.83
C. Transparent-transparent							
	Al_2O_3-SiO_2	100 Al_2O_3	100 Al_2O_3	----	.785	----	.768
	$R^S(SiO_2)$ = 0.034	75-25	64-36	.816	.817	.831	.83
	$R^S(Al_2O_3)$=0.078	50-50	37-63	.840	.817	.880	.83
		25-75	16-84	.858	.85	.919	.903
		100 SiO_2	100 SiO_2	----	.873	----	.95
	ZrO_2-Al_2O_3	100 ZrO_2	100 ZrO_2	----	.75	----	.88
	$R^S(ZrO_2)$=0.14	99-1	98.6-1.4	.751	.75	.878	.88
	$R^S(Al_2O_3)$=0.078	90-10	86.5-13.5	.757	.75	.864	.86
		75-25	68-32	.767	.76	.841	.845
		50-50	42-58	.781	.778	.808	.805
		25-75	19-81	.794	.794	.777	.79
		10-90	7.4-92.6	.801	.81	.762	.79
		1-99	0.7-99.3	.805	.794	.753	.752
		100 Al_2O_3	100 Al_2O_3	----	.805	----	.752

6.0 LIQUID IMPURITIES

It was considered likely that the presence of small amounts of liquid could have a significant effect on the reflectance of powders. Six liquids, namely water, isobutanol, isopropanol, ethyl benzoate, heptane and dibutyl phthalate, were measured out with a pipette, and thoroughly mixed with weighed amounts of powder. Powders used included Al_2O_3, CeO_2, MoO_3, TiO_2, ZnO, CrB_2, $TiCr_2$, Al and Cr. The resultant mixtures were compacted at 11,800 psi into stainless steel holders for reflectance measurements.

In general, with increasing liquid additions, there was a decrease in the spectral reflectance, and the effect was greatest for oxide powders. The decrease in reflectance is illustrated in Fig. 7 for water additions to TiO_2. The reverse effect, that is the increase in reflectance because of loss of water, is included for cases where the powder was heated at 400°C prior to compaction. To be noted are the minima at 1.4, 1.9 and 2.7μ, where water has strong absorption bands. Addition of dibutyl phthalate to CeO_2 results in a similar decrease in reflectance (Fig. 8). In contrast, for powders having opaque particles, the decrease in reflectance because of liquid impurities was much smaller, and fairly independent of wavelength. This is illustrated in Fig. 9 for water additions to $TiCr_2$.

Subsequently, it was realized that the vapor pressure of some of the added liquids was sufficiently high so as to significantly modify the liquid concentration during sample preparation. Thus, the decrease in reflectance caused by adding isopropanol to MoO_3 (Fig. 10) was reversed in about 25 minutes because of volatilization. Also with additions of heptane, water and isopropanol, there were significant changes of reflectance with time. For quantitative studies, liquids having very low vapor pressure (e.g., dibutyl phthalate, vapor pressure $\sim 10^{-3}$ mm.Hg. at 20°C) need to be used.

An explanation of the data needs to consider not only the liquid state of the impurities, but also the larger effects on the reflectance usually observed with oxides than for nonoxides or metals. The oxides have relatively transparent particles, in contrast to the opaque particles of most nonoxides and metals. Therefore, the reflectance of compacted oxides is a volume effect while that for nonoxides and metals is a surface effect. Because of the liquid nature of the additions, the surfaces of the particles are wet by the impurities, resulting in a thin liquid layer spread over the particle surfaces. In the case of opaque particles, the liquid layer may act as an interference film, and thereby decrease the reflectance of the powders in a similar manner to that reported previously for surface oxide films.[5] No absorption

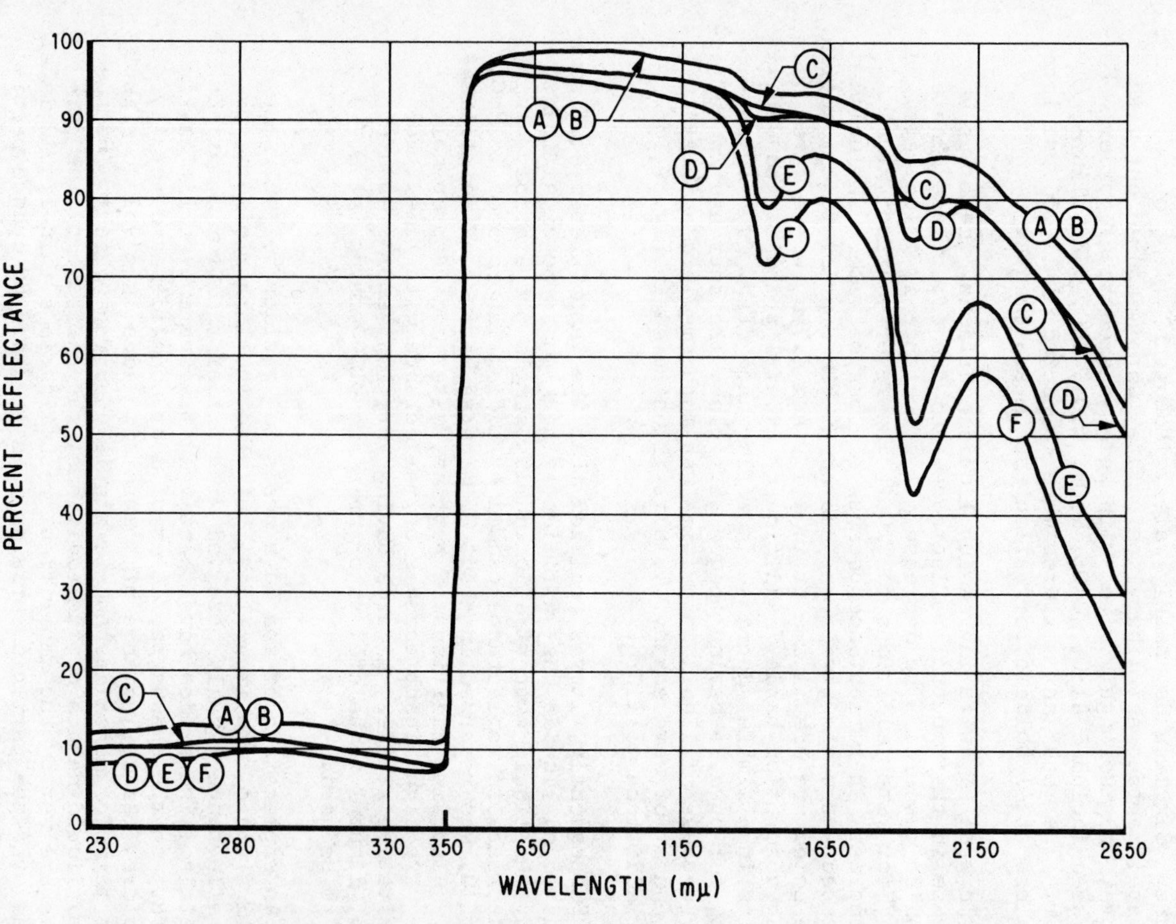

Figure 7. Effect of Water Additions on the Spectral Total Reflectance (vs. MgO) of Compacted TiO_2 Powder: A Heated 24 hrs. at 400°C; B Heated 2 hrs. at 400°C; C As Received; D 2.4% Water; E 4.8% Water; F 9.1% Water. Compacted at 11800 psi.

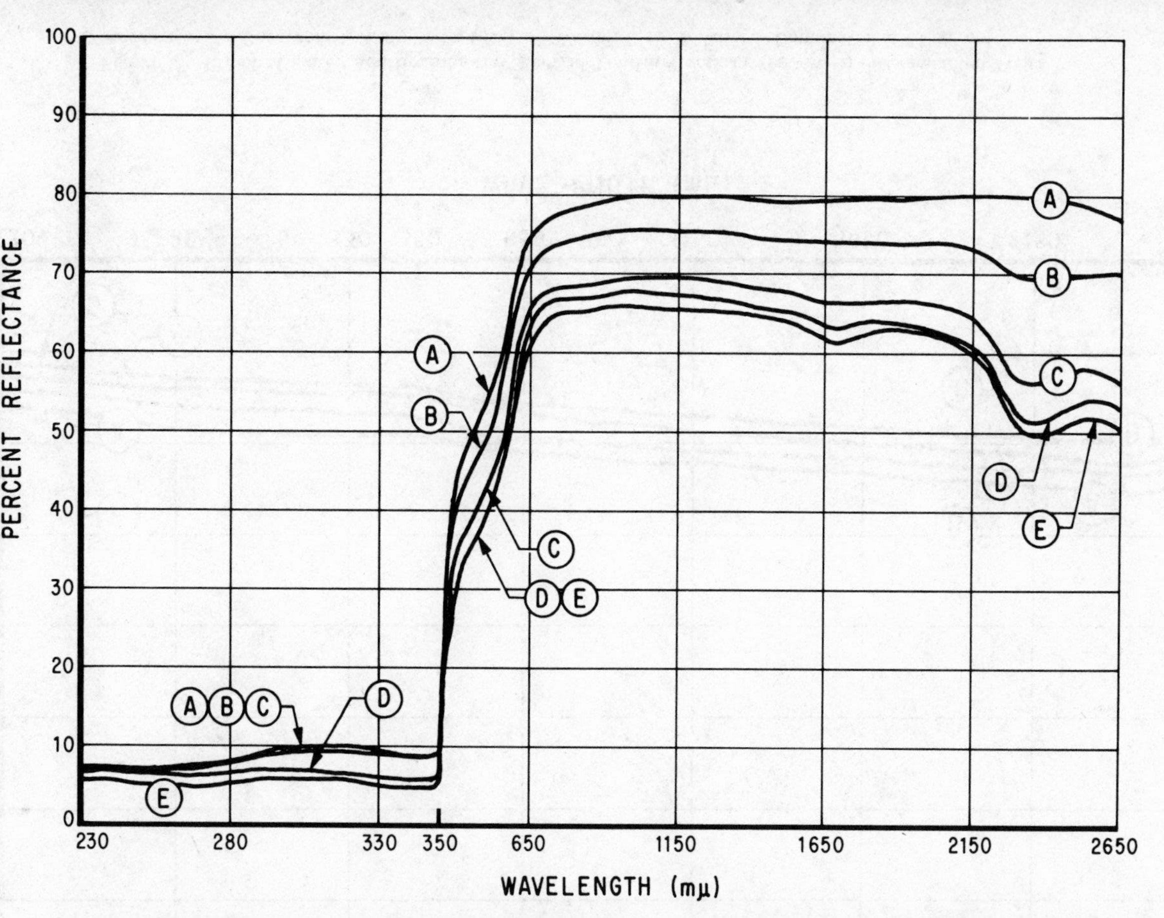

Figure 8. Effect of Dibutyl Phthalate Additions on the Spectral Total Reflectance (vs. MgO) of Compacted CeO_2 Powder. A As Received; B 1.3% Dibutyl Phthalate; C 5.0% Dibutyl Phthalate; D 9.5% Dibutyl Phthalate; E 13.6% Dibutyl Phthalate. Compacted at 11800 psi.

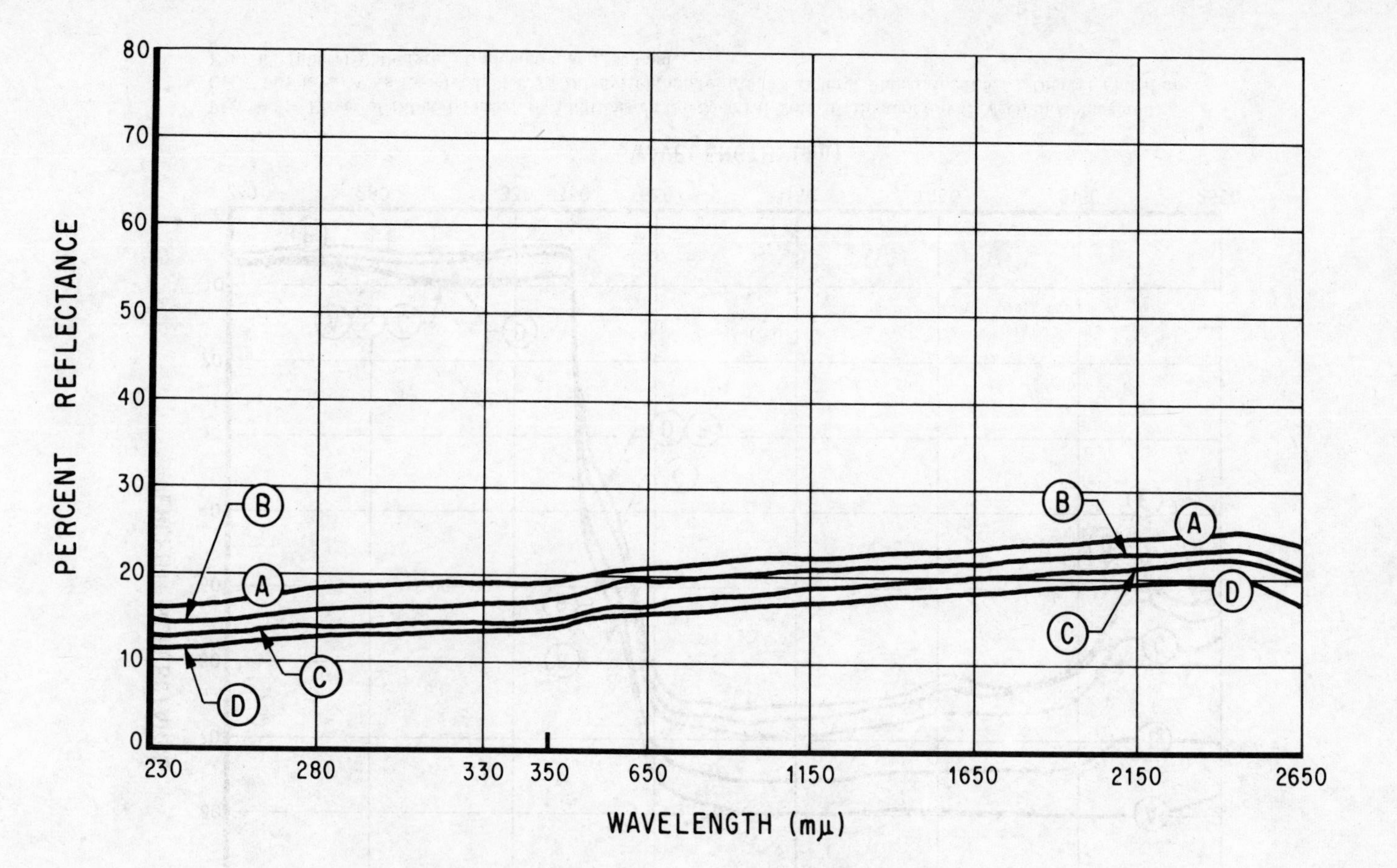

Figure 9. Effect of Water Additions on the Spectral Total Reflectance (vs. MgO) of Compacted $TiCr_2$ Powder: A As Received; B 2.4% Water; C 4.8% Water; D 9.1% Water. Compacted at 11800 psi.

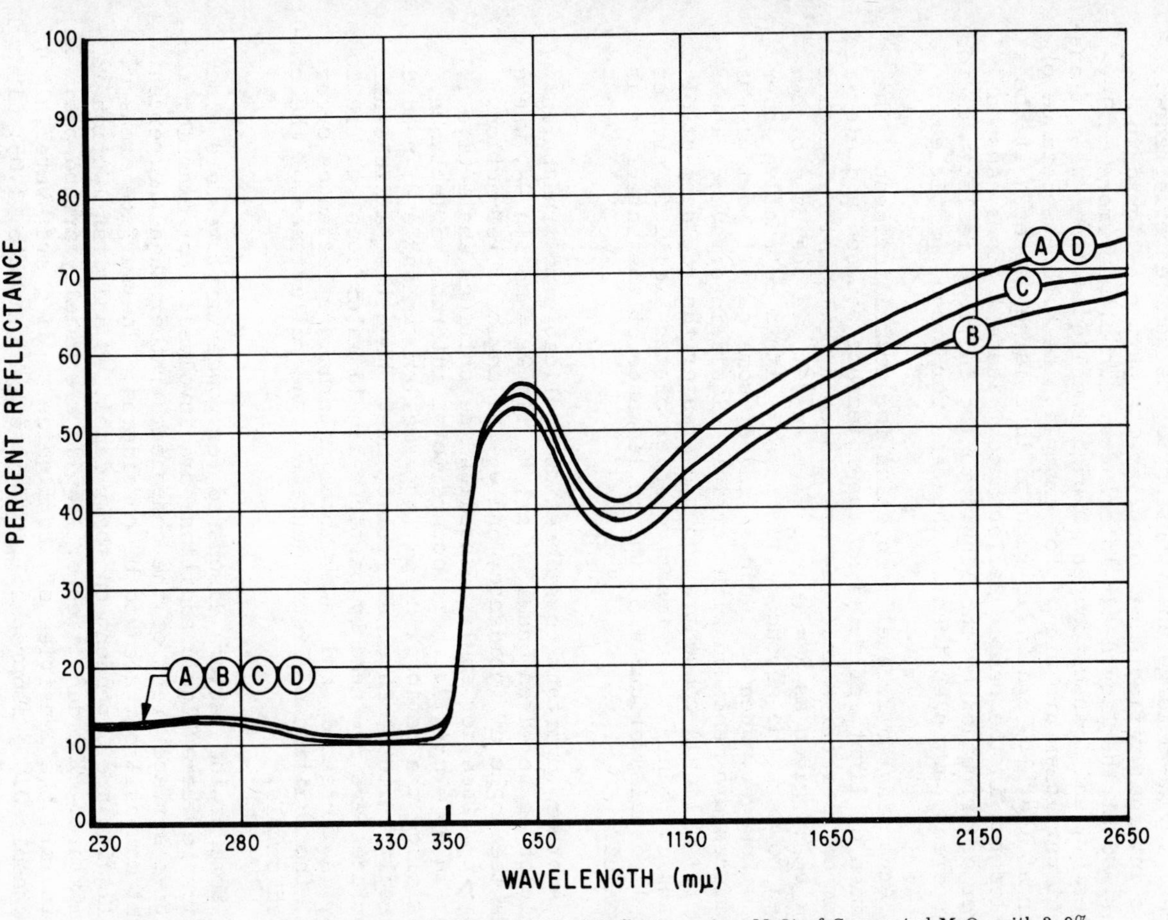

Figure 10. Reversal with Time of the Spectral Total Reflectance (vs. MgO) of Compacted MoO_3 with 3.9% Added Isopropanol: A Pure MoO_3; B After 5 mins.; C After 15 mins.; D After 25 mins. Compacted at 35300 psi.

bands characteristic of the liquid are to be expected except for very thick films.

In the case of transparent particles the internal reflectances provide a large contribution to the sample reflectance. These internal reflectances occur at the particle-liquid interfaces and at the liquid-air interfaces. Thus, many more reflectances by, and transmittances through the liquid occur when transparent particles are used. For example, the high absorptance of water at 1.4, 1.9 and 2.7μ results in a large decrease at these wavelengths in the sample reflectance. These factors cause a larger contribution by the liquid to the infrared reflectance for transparent particles than is the case for opaque particles.

According to Fresnel's equation for normal incidence, the fraction of light reflected at the interface between a dielectric medium of refractive index n_1 and a second medium of refractive index n_2 is given as $R = (n_1-n_2)^2/(n_1+n_2)^2$. In the case of compacted powders the second medium is air with a refractive index of 1. However, when a liquid is added, some of the particle-air interfaces are replaced with particle-liquid interfaces. The higher the liquid concentration the greater the replacement of air by liquid. Since the refractive index of the liquid is greater than 1 (usually around 1.5) the reflectance at the interface is decreased.

Another important consideration applicable to transparent particles is based on Snell's law, $n_1 \sin \theta_1 = n_2 \sin \theta_2$, where θ_1 is the angle of incidence, and θ_2 the angle of refraction. If $n_1 > n_2$, then for angles exceeding the critical angle (i.e., when $\sin \theta_1 = n_2/n_1$), there occurs total internal reflectance. The smaller the ratio of n_2/n_1, the smaller the angle θ_1 needed to result in total internal reflectance. When the particle-air interfaces are replaced by particle-liquid interfaces, the ratio n_2/n_1 is increased, and fewer total internal reflectances occur. Thus, the addition of liquid decreases the reflectance of the compacted sample.

Some of the factors discussed for powder mixtures are also applicable where liquid additions are involved. For example, the relative surface areas of the components rather than the percent weight composition, need to be considered. The surface areas of the liquids are dependent on their densities and distribution in the mixture. The densities of liquids are usually around one. In particular, the densities of the liquids studied are: water, 1.0; isobutanol, 0.80; isopropanol, 0.785; ethyl benzoate, 1.05; dibutyl phthalate, 1.05; and heptane, 0.68. In contrast, the densities of the solid powders used are Al_2O_3 3.97, CeO_2 7.3, TiO_2 4.26, ZnO 5.6, and MoO_3 4.5. Thus, the liquid additions are much less dense than the solids, and therefore their effect on the reflectance of solids are much greater than would be expected based on weight percent. The distribution of the liquids in the mixtures depends on the extent of wetting of the powders by the liquids, and possibly on the particle sizes of the solid components. It seems probable that the liquids surround many of the individual powder particles.

7.0 HEATING EFFECTS

A study was made as to the effect of temperature on the spectral reflectance of compacted oxides. Materials, such as MoO_3, SnO_2, Al_2O_3 and ZnO were compacted and then heated while in the stainless steel holder. Two opposing effects were noted in the case of MoO_3. At temperatures of 100° and 200°C only a decrease in reflectance occurred, which approached a limiting reflectance after about 2 hours of heating. In contrast, heating at 700°C mainly resulted in an increase in reflectance, which was greater when higher compaction pressures were used. At intermediate temperatures, such as 300°C, a decrease in reflectance occurred during the first five minutes, followed by an increase in reflectance upon subsequent heating (Fig. 11).

Heating of compacted SnO_2 also resulted in decreases and increases in reflectance, as illustrated in Fig. 12, although to a lesser extent. Alumina powder, however, only exhibited a slight increase in reflectance at wavelengths longer than 1μ and then only upon heating at 1000°C. Also ZnO only showed an increase in reflectance (Fig. 13).

These results could not be explained as being caused by vaporization of trace amounts of adsorbed water (except possibly for the Al_2O_3) since the effects occurred even when using predried powders. The data can be more readily interpreted in terms of changes in the distances between the particles. At comparatively low temperatures, trapped air may escape resulting in settling of the particles and decreased reflectance. At somewhat higher temperatures the powder may expand upon heating. If some adherence occurs between the particle interfaces, then the distance between the particles will be increased upon cooling. This would result in an increase in the sample reflectance. At still higher temperatures, above those used in this study, actual sintering will occur which would result in a decrease in the distance between the particles and consequently a decrease in the sample reflectance.

Some factors that would determine the temperatures at which these different effects would occur include the coefficient of thermal expansion, the sintering and melting temperatures, the compaction pressure, and the particle size of the powder. Molybdenum trioxide, for example, has a lower melting point (795°C)

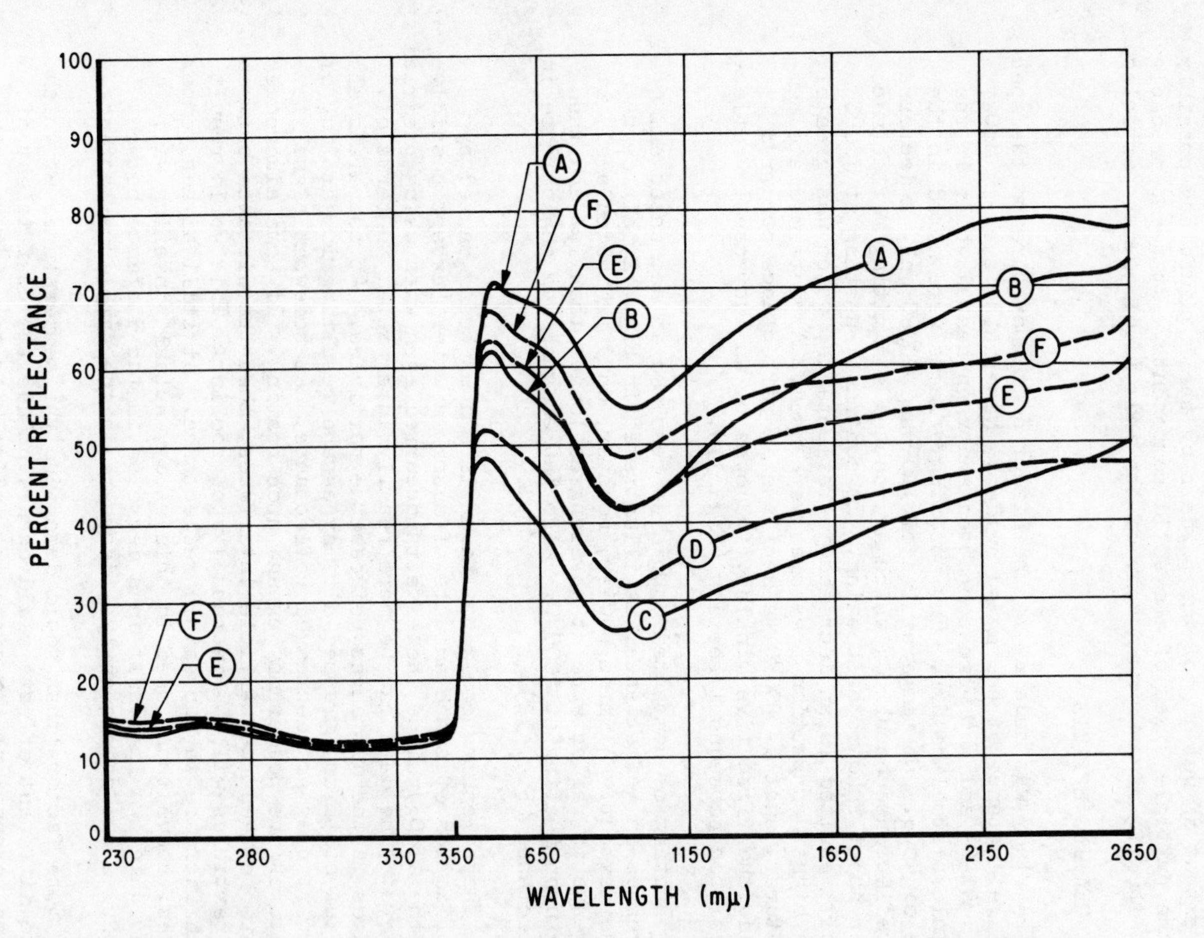

Figure 11. Spectral Total Reflectance (vs. MgO) of Compacted MoO_3 Powders as a Function of Time of Heating at 300°C: Time of Heating (mins.), A 0; B 2; C 5; D 15; E 60; F 150. Compacted Initially at 35300 psi.

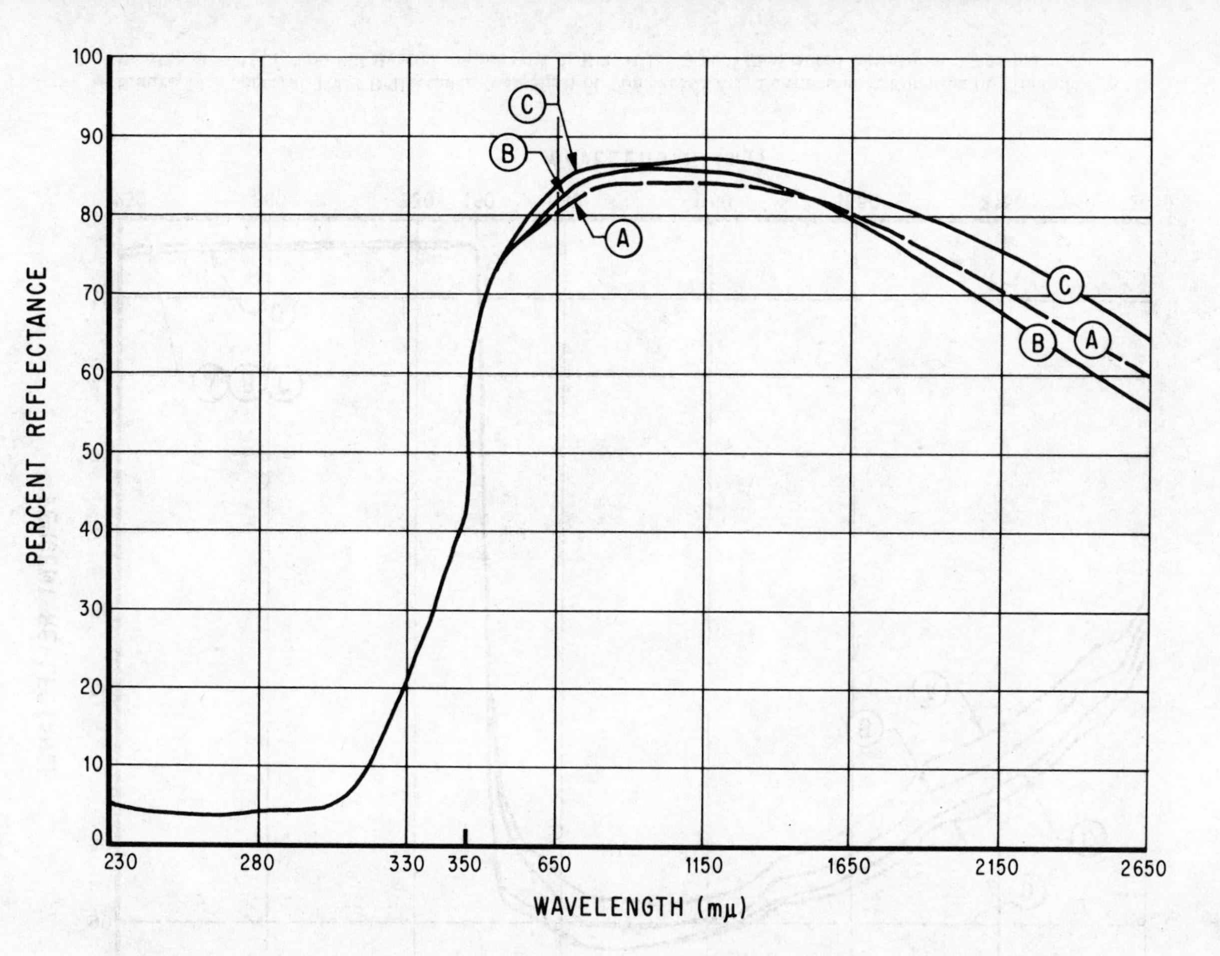

Figure 12. Spectral Total Reflectance (vs. MgO) of Compacted SnO_2 Powders as a Function of Time of Heating at 800°C: Time of Heating (mins.), A 0; B 5; C 60. Compacted Initially at 35300 psi.

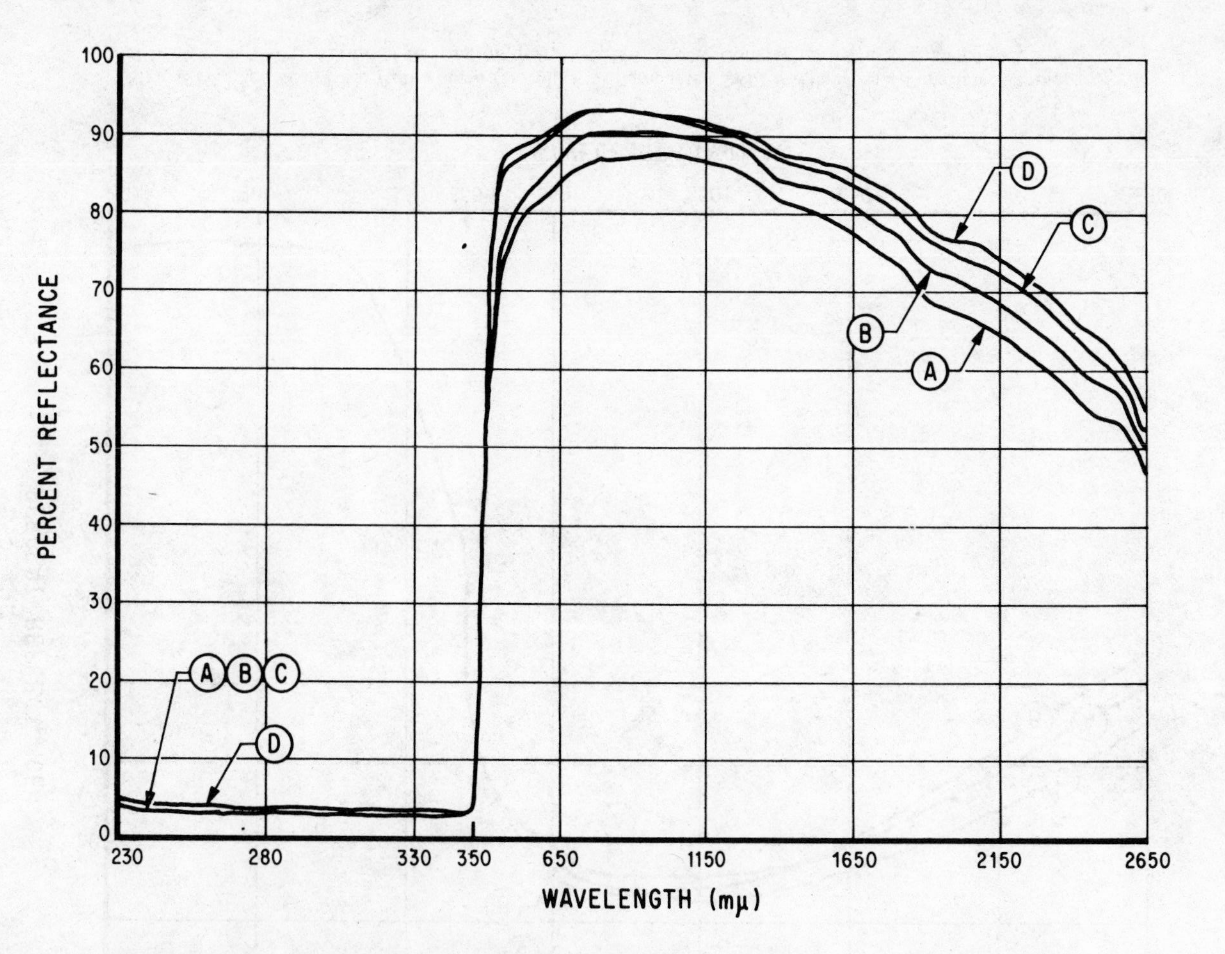

Figure 13. Spectral Total Reflectance (vs. MgO) of Compacted ZnO Powders as a Function of Time of Heating at 500°C: Time of Heating (mins.), A 0; B 2; C 15; D 60. Compacted Initially at 35300 psi.

than the other oxides studied, and therefore the effect of heating is more obvious at lower temperatures. Additional experimentation is, however, needed to insure the correctness of the proposed explanation.

8.0 VACUUM EFFECTS

Research was also conducted to determine the effect of vacuum on the reflectance of compacted powders. Oxide, nonoxide, and metallic powders were studied. In general, somewhat random changes in spectral reflectance occurred. Compacted Al_2O_3 and TiO_2 underwent slight decreases in reflectance with increasing time of exposure, but compacted NiO, SiO_2, MoO_3, Cr_2B, Al and Ni underwent nonsystematic changes in reflectance. A typical result is presented in Fig. 14.

The possibility existed that trace contamination from backstreaming oil could have caused the changes in reflectance. However, upon atomizing diffusion oil onto the surfaces of compacted powders only a decrease in reflectance occurred. Thus, oil contamination could not explain the occasional increases in reflectance that were measured.

It should be realized that the compacted powders not only undergo outgassing upon exposure to vacuum, but also repenetration by air upon removal from the vacuum environment. These effects could cause rearrangement of the powder particles, sometimes resulting in better packing and sometimes in poorer packing. Thus, the extent of compacting of the powders would be expected to change, with a resultant change in spectral reflectance. Whether the extent of compaction is increased or decreased could depend upon such factors as the material, pressure used in sample preparation, and the rate of outgassing and of air repenetration.

9.0 CONCLUSIONS

The presented data indicate that the main variables which control the reflectance of compacted pure powders are the compound used, the opacity of the particles, and the degree of compaction. Other interrelated variables which may also be of importance are particle shape and crystal structure, but these factors are not easily investigated. For example, to study these factors it would be necessary to obtain powders of identical purity, differing only in particle shape or crystal structure. Furthermore, changes in particle shape will also affect the packing of the powder. When considering mixtures, the relative surface area and the relative transparency of the components become of importance. Also, a more complicated situation arises if one of the components is liquid. Parameters that appear to be of importance, only to the extent that they modify the other variables, include temperature, vacuum and particle size.

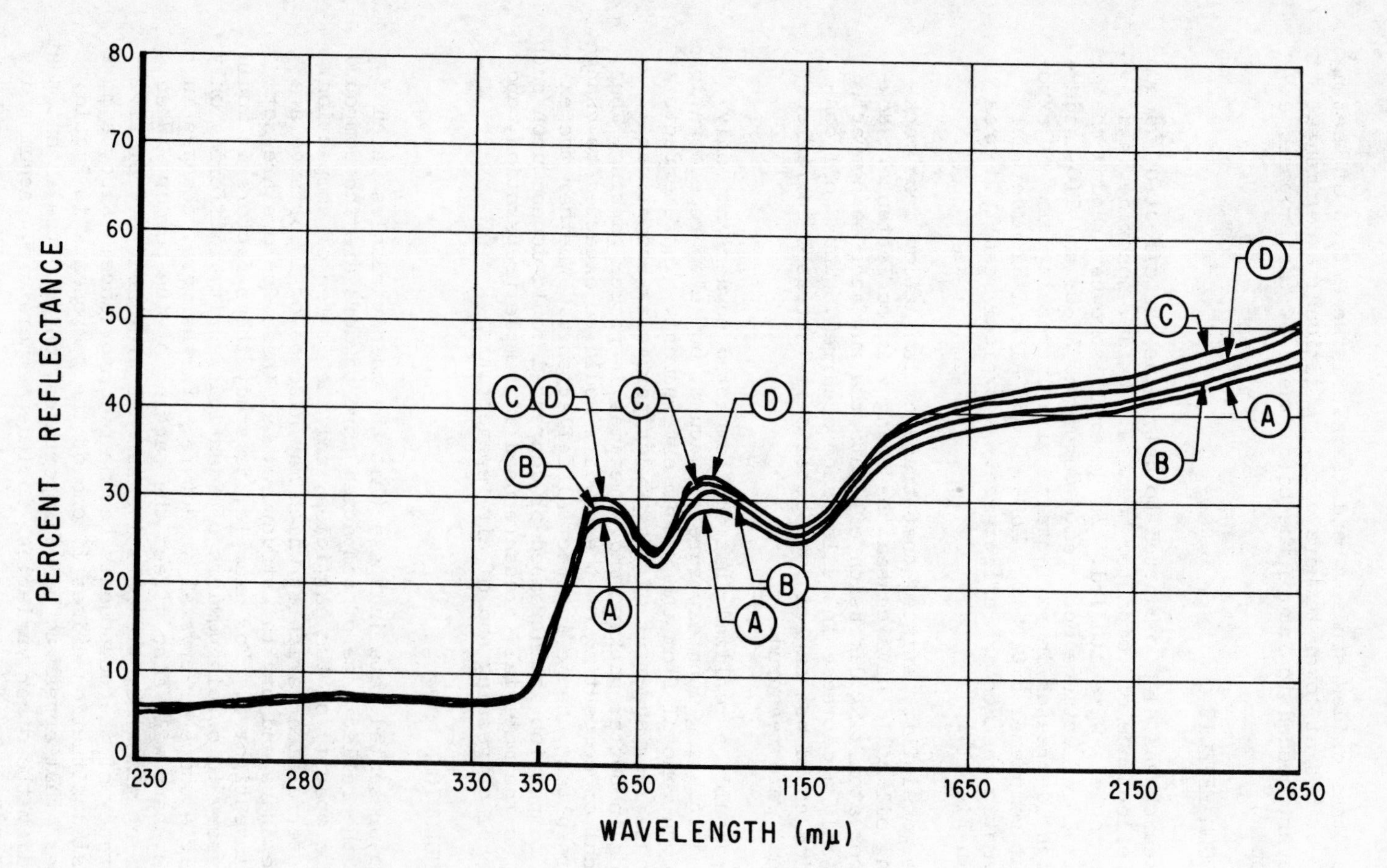

Figure 14. Effect of Vacuum Exposure on the Spectral Total Reflectance (vs. MgO) of Compacted NiO Powder: Time at 10^{-5} Torr (mins.), A 0; B 25; C 105; D 960. Compacted Initially at 23500 psi.

Acknowledgment

This work was performed under U. S. Air Force Contract AF 33(657)-10764, monitored by the Research and Technology Division, Air Force Materials Laboratory, Wright-Patterson Air Force Base, Ohio, with Mr. James H. Weaver acting as project monitor. Appreciation is given to T. L. Burks, M. A. Hoppke and G. Reid who performed most of the experimental measurements.

References

1. A. F. Turner, J. Phys. (Paris) 11:444 (1950).

2. M. Born and E. Wolf, "Principles of Optics", Pergamon Press, New York, (1959) p. 46.

3. E. A. Schatz, J. Opt. Soc. Am. 56:389 (1966).

4. E. A. Schatz, J. Opt. Soc. Am. 57:941 (1967).

5. E. A. Schatz, J. Opt. Soc. Am. 56:465 (1966).

STANDARDIZATION OF MONOCHROMATIC AND ABRIDGED SPECTROPHOTOMETERS

J. A. Van den Akker

Senior Research Associate and Chairman, Department of Physics and Mathematics, The Institute of Paper Chemistry, Appleton, Wisconsin 54911

INTRODUCTION

This paper presents descriptions and discussions of our central laboratory techniques that have been evolved over the years in connection with the operation of an international standardizing system for reflectometers* which have served the double purpose of measuring a specialized reflectance in the pulp and paper industry and functioning as abridged spectrophotometers and colorimeters. In the same period (about 33 years), our laboratories have been the scene of a number of carefully conducted instrumentation studies. In connection with this activity there has also been a need for the development of procedures that would lead to the more accurate and precise use of spectrophotometers and colorimeters.

Before presenting the techniques which we employ, the writer would like briefly to describe our standardizing system and to emphasize the need, in industry, for a color-measuring <u>system</u>, as contrasted with colorimeters furnished with certain standards.

Each of the laboratories participating in our system (a few dozen in 1934, now nearly 500) receives, each month, a parcel of five transfer standards and two working standards, all carefully evaluated on master instruments. Shortly following calibration adjustments, the industrial laboratory reports its results for

*Originally, the General Electric Reflection Meter, and in latter years the Martin Sweets Standard Brightness Tester and Automatic Color-Brightness Tester (AC-BT).

both sets of standards and certain retained working standards. The central laboratory responds to each of these reports with a diagnostic form. If there is difficulty, there is communication, and appropriate corrective steps are taken; most of these can be made in the industrial laboratory but, when necessary, the field instrument is air-expressed to the central laboratory, where correction is effected. On occasion, special standards or materials are sent to the field laboratory to aid in diagnosing whether the seat of the trouble is of a spectral, photometric, or geometric nature.

Before going into the industrial laboratory, each new instrument is sent by the manufacturer to the central laboratory, where it is adjusted with regard to the three classes of instrumental variables: spectral, photometric, and geometric. This involves, for example, the grinding and polishing of filters to achieve both the proper instrumental spectral response and centroid wavelength of that response. The importance of the latter is often overlooked in industrial colorimetry: While accuracy of a colorimetric quantity is of primary importance, good agreement between two or more instruments in the same or different companies is a necessity, in view of commercial and governmental specifications as well as proper maintenance of technical control.

WAVELENGTH STANDARDIZATION

Spectrophotometers

The Hardy-General Electric Recording Spectrophotometer (GERS) has always been the mainstay of our spectral standardization and also a valuable aid to our photometric work, first in the model employing a chopper, and later (1937) in the form which, in basic principle, exists today. Until 1939 we had used the several means which were then (and now!) available for correcting the wavelength scale of the GERS. In that year, dissatisfied with the best of those methods, we developed a procedure for the direct measurement of the effective wavelength (centroid or first moment) of the spectral response of the instrument. With minor procedural improvements, we have used the method[1] with good success during the intervening period of 28 years.

It is especially important, in our opinion, to employ the method where the wavelength bandwidth is 5 or 10 nm, or more, as this range is large in comparison with the desired accuracy, and the radiation is to be considered heterogeneous rather than monochromatic. Over much of a spectral reflectance recording, the curve is nearly straight and, where it is straight, effective wavelength has exact significance, irrespective of the algebraic sign or magnitude of the slope of the curve. Hence, the use of effective wavelength is meaningful. The concept has the additional advantage that it can be determined directly, with good accuracy (to about 0.2 nm or better), the level of accuracy depending entirely on the care exercised and the number of replications of the measurement.

The interested reader is referred to the original paper on the application of the method to spectrophotometers[1]. Since the appearance of that publication, further spectrum lines have been added through the use of a cadmium lamp. Only a few additional remarks need to be made here. First, it is important to note that the "spectrum-line transmittance" of a linear filter ($\underline{T} = \underline{a} + \underline{b}\lambda$ over the wavelength bandwidth) should preferably be determined by means of the instrument under calibration, with the instrument actually operating on spectrum light, because errors resulting from photometric inaccuracy and changing geometric conditions (e.g., angles of rays, portion of filter illuminated) are then negligible. Second, the slope of the spectral transmittance curve of each filter should be large in the wavelength range for which the filter is selected. Third, the fluorescence of the filter, when under illumination at the wavelength of interest, should be negligible. Fourth, when the spectrum light is comprised of two or more closely neighboring lines, the properly weighted mean wavelength of the group should be determined by suitable spectroradiometric equipment. In this special determination, the spectrum light should be that which has passed through the spectrophotometer under conditions simulating normal operation. Finally, if the monochromatic light source with which the GERS is operated exhibits a "ripple" when the power supply is a d.c. generator, the instrument may display an objectionable hunting. This is easily eliminated through the use of a conventional inductance-capacitance circuit.

The principle of the method can be used in laboratories that are staffed only with laboratory assistants. The spectrum-line transmittance of each of a set of linear filters (chosen to yield both positive and negative slopes at each wavelength) is carefully determined in a central laboratory. Depending on the degree of dissimilarity of the spectrophotometers in the central and field laboratories, it may, or may not, be desirable to make the determinations with suitably chosen clear blanks in the reference beam. Photometric accuracy of both the central laboratory and field instruments is now a factor — but a factor that is easily eliminated. The photometric correction of the central laboratory instrument is known, and the spectrum transmittances of the filters are corrected. If the photometric error of the field instrument (at a particular value of $\underline{T}$) is negligible, the indicated wavelengths obtained with the pair of linear filters of positive and negative slope for a particular true wavelength will be in agreement; that is, the λ_1 read from the instrument when the wavelength is manually adjusted to yield the spectrum transmittance $\underline{T}_1$ of the filter having a positive value $\underline{b}_1$ (see foregoing paragraph) will be equal to the λ_2 obtained with the other filter having a negative slope $\underline{b}_2$. If the photometric error is not negligible, the pair of indicated wavelengths will differ, and it is easily shown that the indicated wavelength, corrected for photometric error, can be computed by means of Eq. (1).

$$\lambda' = (\underline{b}_1 \lambda_1 + \underline{b}_2 \lambda_2)/(\underline{b}_1 + \underline{b}_2). \tag{1}$$

In Eq. (1), $\underline{b}_2$ is the magnitude of the slope of the spectral transmittance curve of the linear filter having negative slope at the check point [the negative sign has been taken into account in arriving at Eq. (1)]. The equation is most accurate when the two $\underline{T}$-values are nearly the same (so that the photometric errors in the two determinations are the same) and, hence, the pairs of linear filters are chosen to have similar transmittance at the check points.

The correction to be added to the wavelength scale of the instrument under calibration is, of course, the difference between the known wavelength for the check point and λ'.

If the temperature coefficient of a filter is significant, a temperature correction must be made.

The method furnishes an auxiliary check on the photometric accuracy of the instrument under calibration. The error to be added to the photometric scale is

$$\underline{e} = (\lambda_1 - \lambda_2)\underline{b}_1\underline{b}_2/(\underline{b}_1 + \underline{b}_2), \tag{2}$$

in which the units of $\underline{e}$ are determined by the choice of units for $\underline{T}$ (and the $\underline{b}$-values).

Abridged Spectrophotometers

In connection with both our standardizing work and instrumentation program, we[2-11] frequently utilize the method presented above for the determination of effective wavelengths of filter-type instruments (and, additionally, the instrumental spectral response for the several filters of tristimulus colorimeters).

Where the wavelength range is large, it is obviously more difficult to obtain filters that are adequately linear over the active range. In some of our earliest work(2), good results were obtained with liquid filters (e.g., degassed, properly diluted Coca-Cola is almost exactly linear over the range 400-500 nm). Later, colored glass and glass filters were found which yielded good results. In some cases, especially where the wavelength range is quite large, no material has been located which is sufficiently linear. In this situation, the best available material is used, and a correction is applied. The correction is obtained very precisely and straightforwardly through the use of an analogue computer(12). The computer is set up to simulate the operation of measuring effective wavelength, first with a simulation of a perfectly linear filter (the other function of wavelength being the instrumental spectral response), and then with the spectral transmittance of the actual, imperfectly linear filter fed into the integration. The nature of the correction (which is small) is such that this process need not be repeated for each instrumental calibration.

It is usually possible to locate a point in the optical train of a reflectometer at which a legitimate measurement of transmittance can be made. Certain precautions relating to geometrical considerations must be observed. Since the presence of a filter alters the optical path length, the measurement of transmittance, in both the filter instrument and spectrophotometer, should embrace a nonabsorbing glass (or liquid filter) of equivalent optical thickness. If the range of angles of the rays in the filter instrument is substantially greater than the spread in the spectrophotometer, it is often legitimate to arrange an *ad hoc* narrowing of the range or, if inspection suggests that this should not be done, a correction to the transmittance observed with the filter instrument can be computed.

As in the wavelength calibration of a spectrophotometer in the field, linear filter combinations of both positive and negative slope are employed, and Eq. (1) is called into play. There is an obvious procedural difference: For a given filter of an abridged spectrophotometer or colorimeter, the pair of transmittance values are referred to the corresponding, corrected spectral transmittance recordings (in our case, GERS recordings), and the values of λ_1 and λ_2 are read off. Equation (1) is then used to obtain the effective wavelength, and Eq. (2) is employed if the auxiliary check on photometric accuracy is desired. Where the spectrophotometer is in the same laboratory, a preferred procedure is to place the filter combinations in the spectrophotometer, manually adjust the wavelength to obtain the same transmittance, and then apply the wavelength correction. Again, for filters having significant temperature coefficient, correction must be made for temperature difference.

Materials having linear reflectance ($\underline{R} = \underline{a} + \underline{b}\lambda$) over the wavelength range of interest have been employed for diagnostic purposes, with special reference to the spectral aspects of an instrument in a field laboratory. However, there are obvious pitfalls. The most serious of these are errors arising in the differences in geometry of the standardizing and field instruments. The reflectance of all the materials that we have examined over the years has depended on the geometry of illuminating and viewing. This means, for example, that the spectral reflectance determined with a spectrophotometer equipped with an integrating cavity will be different from that determined with another instrument (spectrophotometer or filter-type) employing the CIE 45°-0° mode of illuminating and viewing. In addition, there are other factors of geometric origin that can account for serious errors in attempts to evaluate effective wavelength through the use of materials having linear reflectance.

PHOTOMETRIC STANDARDIZATION

Where a spectrophotometer of known photometric correction is available, the photometric calibration of other instruments is obviously simplified. Accordingly, we initiated efforts to test the photometer of our first polarization-type GERS during its first months in our laboratory (1937). Linearity was checked by a number of well-known methods. For example, the transmittance of lightly sooted screens, chosen for uniformity, was measured for the screens taken singly and in series combination (with coarse, moiré pattern obliterated by suitable adjustment of mutual angle about the optic axis of the instrument). In another effort, the tangent-squared cam was tested by measuring the displacement of the cam follower by means of the precision screw from a Michelson interferometer. This showed that the grinding of the cam was nearly perfect. This technique does not test the photometric accuracy, as there are other links in the chain. But it is of good diagnostic value; for example, the data clearly revealed that the hole in the tiny roller on the cam follower was slightly off center!

It soon became clear to us that the nonuniformities of screens (even with controlled active area) were responsible for intolerably large error. In our next step, a transmittance standard was made by boring and bevelling holes of known spacing and diameter (finished with a special tool) in a sheet of metal. The transmittance of this standard for parallel rays and rays of known divergence was calculated. To be sure, this standard furnished us with only one check point (just under $\underline{T} = 0.5$) — but this point together with the demonstrated accuracy of the tangent-squared cam assured us that the photometric accuracy of our spectrophotometer was excellent. This transmittance standard has proved to be very useful in a number of different studies.

There remained, however, a nagging concern about an unexpected or unknown source of photometric error in the system at points away from the check point. Approximately twenty years ago, Mr. Dearth of our laboratory developed an adaptation of the Wood(13) zero-resistance circuit for calibrating the whole photometric scale of spectrophotometers, filter-type instruments, opacimeters, and transparency meters. (The electron current in this circuit is linearly related to light flux.) The following is a brief description of the method as used for checking the photometric accuracy of the GERS.

The zero-resistance circuit contains a selected Weston blocking layer photocell and a high-sensitivity galvanometer (1.5×10^{-9} amp/div). The deviations from linearity of the galvanometer (determined by means of the circuit described in any textbook on electrical measurements) are accurately known. The circuit contains a double-pole-double-throw switch, so that the galvanometer can be used both as the null-point detector and current indicator. The system is checked for linearity at a single point near midscale by means of the standard transmittance referred to earlier; this is done in our master transparency meter, which provides a very steady beam in which the angular spread of the light rays is small, and known. (Generally, if a photometric system of this sort is nonlinear, the greatest divergence from linearity is found near midscale.) With the GERS _operating in normal manner_, the instrument is balanced at 100.0 with two matched surfaces of high reflectance. The method is such that two matched piles of paper (translucent, not opaque) of reflectance higher than 0.9 are placed over the standard and specimen ports of the GERS. The number of sheets used in each pile is such that full-scale deflection of the galvanometer can be obtained when the photocell is placed over the pile covering the specimen port (the photocurrent arises in the low-level transmitted light). The pile covering the standard port is backed with a black body. A wavelength for which the response of the photocell is maximum is chosen for the determination. After balancing both the GERS and the galvanometer deflection to 100.0, the first of a series of neutral screens of appropriate transmittance is placed in the specimen beam and nearly simultaneous readings are taken from the GERS counter and galvanometer deflection. This is repeated until the desired number of check points across the scale has been obtained. Of course, the galvanometer correction is applied at each point. Careful use of this method, with several replications at each point, yields an accuracy estimated to be about 0.1% of full scale. The reader will note that uniformity of the screens, which act simply as fixed attenuators, is not a factor influencing the accuracy of the method; in fact, in the separate uses of given screens, significantly different transmittances are observed, but the corrections remain the same.

Dearth's method is employed in the calibration of both double-beam and single-beam instruments. In each case, the zero-resistance circuit operates on the low-intensity light transmitted through a pile of paper sheets that is presented to the specimen port of the instrument. Although the pile is sufficiently translucent for the determination, its reflectance and reflectivity ($\underline{R}_\infty$) are almost identical.

GEOMETRY

Of the whole array of factors influencing the overall accuracy and precision of spectrophotometers and filter-type instruments, some of the more important (but often overlooked) instrumental parameters are classified as "geometric." Of these, the best known relate to the "geometry of illuminating and viewing." Different geometries (e.g., those of CIE 45°-0° and the integrating cavity) generally yield significantly different reflectance. Our present concern is that of standardization.

For the CIE 45°-0° mode of illuminating and viewing employed in the instruments standardized by our system, the following parameters are recognized and controlled. Long experience has shown that the axes of the illuminating and viewed beams must be kept well within 0.5° of 45° and 0°, respectively. In so far as the reflectance of various papers is concerned, this requirement is far more stringent than the angular spread of the rays in the two beams, although the latter is controlled.

In the geometric category is "edge effect," which we have usually termed the "translucency effect." The effect, which is well recognized, is a complicated function of illuminated area, viewed area, associated linear dimensions, scattering coefficient and thickness of the specimen, and perhaps other factors. It is subject to control, although the reference for the effect has lain, so far, in standard materials. It would seem that the best ultimate reference that one can conjure is one involving directional reflectance for infinite areas of uniform illuminating and viewing, although it can be argued that the same result would be obtained for the case of uniform illumination over an infinite area, with viewing of a finite area. We have made some attempts, using the GERS, to evolve an extrapolative procedure in which the viewed area was varied; the work has been particularly useful in the control of reflectance of such translucent material as creped facial tissue.

On considering the geometric parameters of the integrating cavity mode of illuminating and viewing, one's first thoughts naturally turn to the question of including or excluding the "specular component." In many diffusely-reflecting materials (paper is a common example), there is no definable specular component, for the lobe or spike on the goniophotometric surface is broad in some materials, narrow in others, and in all cases gradually joins the main body of the surface. A much less recognized but nevertheless very important factor of the integrating cavity geometry is the absolute reflectance of the cavity lining, the area of the lining, and the fraction of the total cavity area occupied by ports. In the case of some field laboratory instruments which we have seen, the reflectance of the "white" lining had deteriorated to the point where it would be difficult to say whether the geometry of the instrument was closer to that of the CIE 45°-0° system or the ideal integrating cavity. A number of years ago, Mr. Dearth made some calculations that indicated that, if the influence of the first-reflected flux from the specimen were to be negligible*, the absolute reflectance of the lining — measured by means of equipment that is not to be confused with that to which reference is made later(14) — should be higher than 0.955. This level, then, has been our arbitrary lower limit where we use a GERS equipped with an integrating cavity.

REFLECTANCE STANDARDIZATION

For a number of years the ultimate standard of reflectance in our laboratories, as in many others, was MgO. As in other laboratories, the impact of the variability of MgO had been minimized through the use of well-aged, well-protected materials of nearly constant reflectance. We encountered an intolerable situation: a newcomer to our laboratories consistently produced MgO surfaces of reflectance about 1% higher than that which could be achieved by two older, skillful men. The three men studied their techniques carefully, but there was nothing that the older men could do to diminish the 1% difference. It was then that we returned to an earlier search for a method which, in the hands of laboratory assistants, would yield accurate data on the absolute reflectance of diffusely-reflecting media like MgO and $BaSO_4$. The development of a suitable method, described elsewhere(14), was concluded in 1950 and, after a trial period of two years, was put into use in connection with our international standardizing system. Goebel, Caldwell and Hammond(15) have carefully studied the method and made an improvement which we are currently using with good success.

*The first-reflected flux corresponds to a unidirectional reflectance at approximately (6°, 52°).

Looking to the future, improved reflectance standardization and instrumentation both require a better understanding of the dependence of the reflectance of diffusely-reflecting materials (standards and specimens alike) on the corresponding goniophotometric surfaces. In the case of directional reflectance, proper numerical integration of goniophotometric data over the angular ranges of the illuminating and viewing beams is fundamentally involved. Integration, for widely different limits, is also involved in reflectance as determined with an integrating cavity instrument. Actual goniophotometric surfaces depart widely from the sphere of the ideal diffuser; as is fairly well known, the departure even for MgO (as to shape, not magnitude) is quite significant. To obtain more meaningful data for the understanding of the gloss of paper(16) and reflectance of materials generally, Dr. Scheie of our laboratory has recently designed and directed the construction of a goniophotometer to replace our old research model.

International standardizing organizations have already established a schedule for the replacement of MgO by absolute reflectance as the controlling standard. One can well understand the concern of industrial people using the integrating cavity geometry in view of the fact that this change in ultimate standard will change their reflectance scales by approximately 2%. The situation is different for laboratories using the CIE 45°-0° geometry; the work of Preston(17) has shown that, for this mode of illuminating and viewing, the directional reflectance of MgO is the same as that of the ideal diffuse reflector (apparently an accident of nature). However, the MgO used over the years (before 1950-52) in our laboratory may not be quite the same as that employed by Preston and, hence, Mr. Dearth and the writer are planning a very careful goniophotometric study of the MgO standard which we have employed, utilizing our method for determining absolute reflectance.

LITERATURE CITED

1. J. A. Van den Akker, J. Opt. Soc. Am. 33:257 (1943).

2. J. A. Van den Akker, Philip Nolan, and W. A. Wink, Paper Trade J. 114(5):34 (1942).

3. L. R. Dearth, J. A. Van den Akker, and W. A. Wink, Tappi 33:85A (1950).

4. J. A. Van den Akker, M. Aprison, and O. H. Olson, Tappi 34:143A (1951).

5. L. R. Dearth, W. M. Shillcox, and J. A. Van den Akker, Tappi 34:126A (1951).

6. J. A. Van den Akker, L. R. Dearth, O. H. Olson, and W. M. Shillcox, Tappi 35:141A (1952).

7. L. R. Dearth, W. M. Shillcox, and J. A. Van den Akker, Tappi 41:196A (1958).

8. L. R. Dearth, W. M. Shillcox, and J. A. Van den Akker, Tappi 43:230A (1960).

9. L. R. Dearth, W. M. Shillcox, and J. A. Van den Akker, Tappi 43:253A (1960).

10. L. R. Dearth, W. M. Shillcox, and J. A. Van den Akker, Tappi 46:179A (1963).

11. L. R. Dearth, W. M. Shillcox, and J. A. Van den Akker, Tappi 50(2):51A (1967).

12. J. A. Van den Akker, J. Opt. Soc. Am. 29:364 (1939).

13. L. A. Wood, Rev. Sci. Instr. 7(3):157 (1936).

14. J. A. Van den Akker, Leonard R. Dearth, and W. M. Shillcox, J. Opt. Soc. Am. 56:250 (1966).

15. D. G. Goebel, B. P. Caldwell, and H. K. Hammond, III. J. Opt. Soc. Am. 56:783 (1966).

16. J. A. Van den Akker, and G. R. Sears, Tappi 47:179A (1964).

17. J. S. Preston, Trans. Opt. Soc. (London) 31:15 (1929-1930).

INSTRUMENTS FOR REFLECTANCE SPECTROSCOPY

Werner E. Degenhard

Carl Zeiss, Incorporated

Introduction

The measurement of reflectance is one of the most difficult problems of spectroscopy. Reflectance is not a simple specific property of a sample as is absorbance or polarisation. Reflected light is influenced by many other parameters such as illuminating light, illuminating aperture, surface texture of the sample, polarisation and absorbance of the sample. The light can be reflected diffusely, specularly or as a mixture of both. The samples can be flat, curved or irregularly shaped.

The human eye is the most perfect instrument with which to determine reflectance regardless of the property of the sample, but the sensation or impression received by the human brain cannot be converted into physical values. Reflectance cannot be separated from other parameters by a single instrument. We are forced to simplify and to standardize. The instrument designer has to rely on international agreements concerning illumination, "standard observer" and measuring geometry. He must consider the different types of samples: powders, pigments and fine crystals, textured surfaces such as those of textiles, highly glossy surfaces such as metals and paint, and, as a special task, to measure very tiny samples.

Finally he is now encountering more and more samples which are not only colored but also fluorescent.

1. Instrumentation

a) Wavelength selection. A general differentiation can be made regarding the measuring procedures. One type of instrument measures reflectance manually or automatically, wavelength after wavelength throughout the visible range of the spectrum (usually 380 to 720 nm). These are the Reflectance Spectrophotometers. For some particular purposes the range can be extended into the ultraviolet or infrared region. Wavelength selection in this type of instrument is made by a monochromator.

The other type of instrument selects the wavelength by means of filters: the Filter-Photometers. The difference between these two is obvious: spectrophotometers are more accurate because they measure at many more points of the wavelength scale and the resolving power is greater than that of a filter. But they are more expensive and the operation is time consuming. They are hardly suitable for the application in a workshop or mill. Filter-Photometers are less accurate but less expensive and faster in operation. They are sturdy and better suited for the daily routine job.

b) Measuring geometries. Both instrument types can be equipped with a desired measuring geometry. Two geometries are internationally standardized: illumination under 45 degrees and observation under 0 degrees (or vice versa) and diffuse illumination

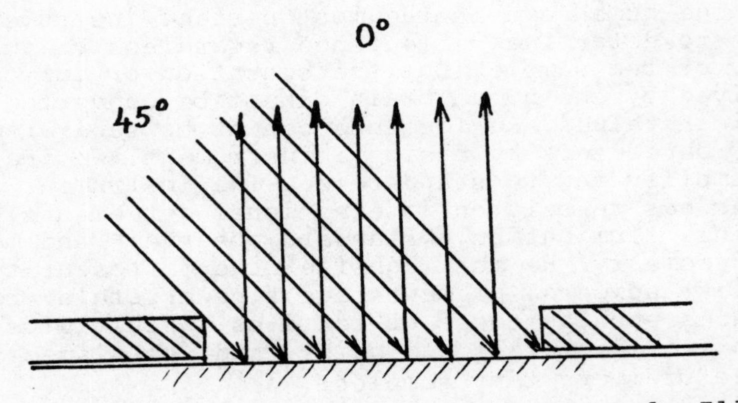

Fig. 1. Reflectance of an opaque gloss sample Illumination 45°, observation 0°.

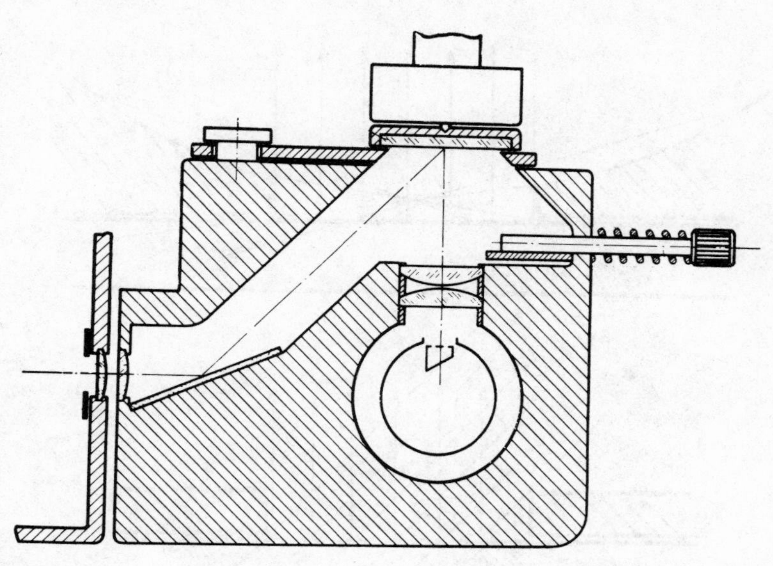

Fig. 2. Cross section of a reflectance attachment with measuring geometry 45°/0°.

and observation under 0 degrees (or vice versa). The abbreviations are 45°/ 0 and d/0. The following pictures show some typical effects on some samples which of course can largely vary.

Fig. 1 shows a completely opaque highly glossy sample such as a metal surface. The measuring geometry is 45°/0. Gloss is completely eliminated.

Fig. 2 shows the cross section of such a design, in this case an attachment to a manually operated spectrophotometer.

When the sample is not completely opaque, however, there is some doubt about the validity of the readings. This may be seen from Fig. 3.

The light which enters the left portion of the sample is not illuminated and does not contribute to the amount of light reflected back. This example applies to most of the materials in industry such as pigments, powders and paper as well as to all sorts of fibers. Also the widely used standards: magnesium oxide, barium sulphate and titanium dioxide are themselves more or less translucent. A layer of MgO-powder has to be 2mm thick before the influence of the background disappears.

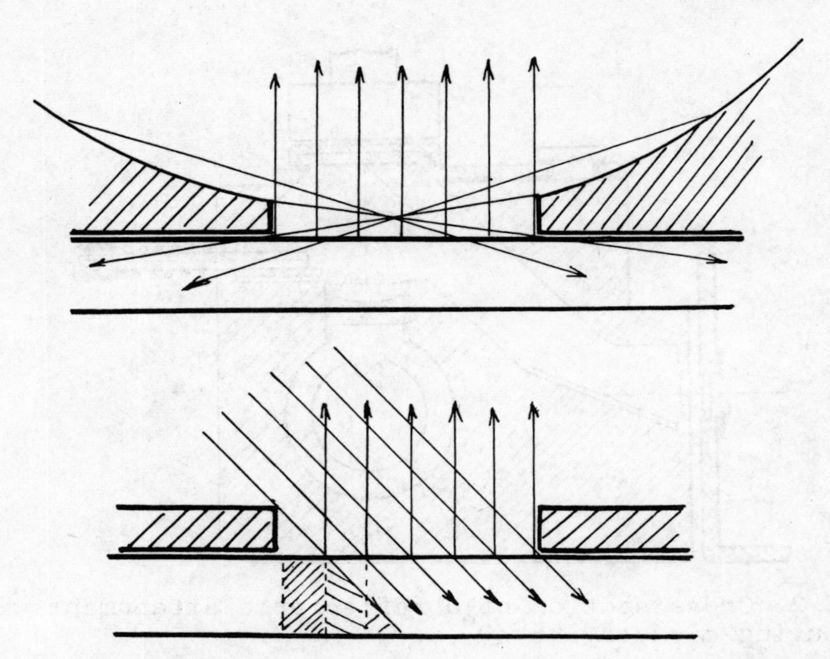

Fig. 3. The effect of translucency. Comparison of 45°/0° and sphere geometry.

With diffuse illumination however all parts of the sample are illuminated and all parts are able to contribute to the back scattered light. Another handicap of the 45° illumination is the effect of surface structure of the sample. Fig. 4 shows the result somewhat exaggerated.

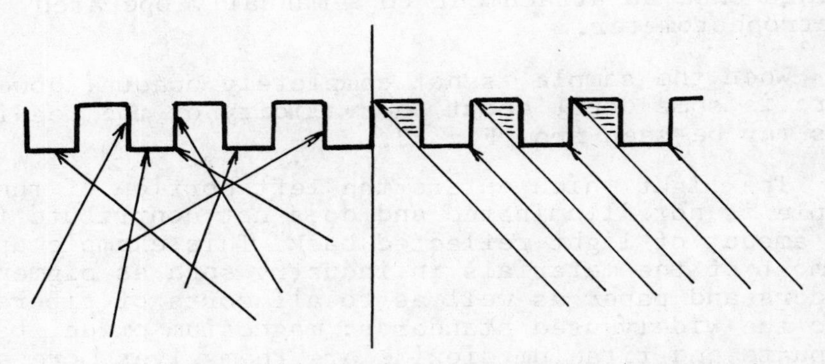

Fig. 4. The effect of surface texture. Comparison of 45°/0° and sphere geometry.

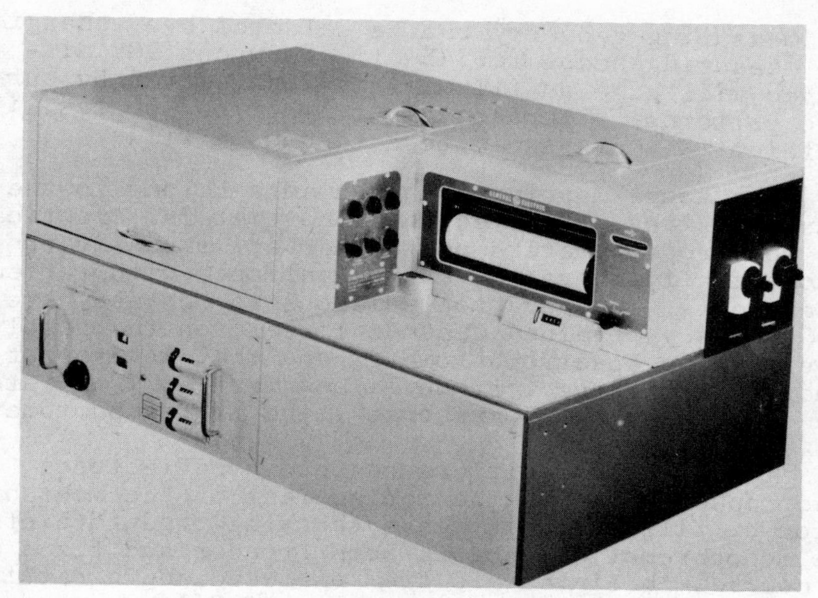

Fig. 5. General Electric Recording Spectrophotometer

It is obvious that the dark portions of the sample area do not participate in the reflectance. The readings are too low. This effect can also be eliminated by the d/0 geometry, in other words, by the application of a photometric sphere. If the surface is illuminated from all directions by the inner surface of the white coated hollow sphere the readings should give true values of the diffusely reflected light. By screening the exit pupil of the sphere, gloss can be totally excluded.

2. Instruments for Reflectance Measurement

The standard instrument in this country which is considered to give the most accurate values is a recording spectrophotometer with a sphere geometry. Fig. 5. As a double beam instrument it records the ratio of reflectance of a standard and the sample. It can be equipped with an integrator which displays the tristimulus values in digits.

A German version is shown in Fig. 6. It incorporates a double monochromator and an integrator for direct evaluation of the tristimulus values.

The measuring geometry can be selected by exchanging the measuring head. Two 45o lightsources 90° displayed will give an illumination unaffected by surface patterns. Alternatively two hollow spheres with different diameters can be attached at will.

The above mentioned instruments belong to the "luxury" class of reflectance photometers. Practically all spectrophotometers on the market, manually or automatically operated, can be equipped with reflectance attachments. They are too numerous to mention here. The accuracy obtained depends largely on the resolving power of the spectrophotometer and the diameter of the sphere. It will be worthwhile,however, to look into the details of the "workhorse": the filter photometer.

The principle difference from a reflectance spectrophotometer is the replacement of the monochromator by filters. This means that the bandwidth of the monochromatic light is both broader and fixed and, of course, the intervals from one wavelength setting to another are larger. The number of filters used in a given instrument varies widely. The main advantage of a filter photometer, beside the ruggedness and easy operation, is that the standard light source and the standard observer can be incorporated in the filters. The filter instrument is then able to give the tristimulus values directly without manual or electronic computation of the characteristic reflectance of the sample. The accuracy is slightly lower than that obtained by a spectrophotometer but it is almost within the threshold values of the human eye. Both of the previously mentioned measuring geometries are applied in present instruments, but if the same sample is

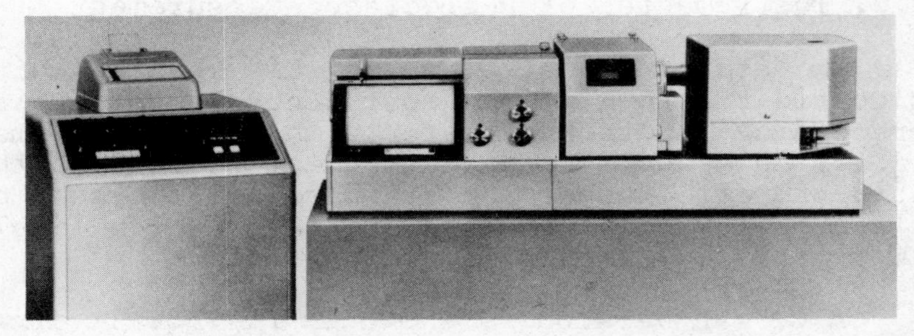

Fig. 6. Carl Zeiss Recording Spectrophotometer for Color Measurements DMC 25.

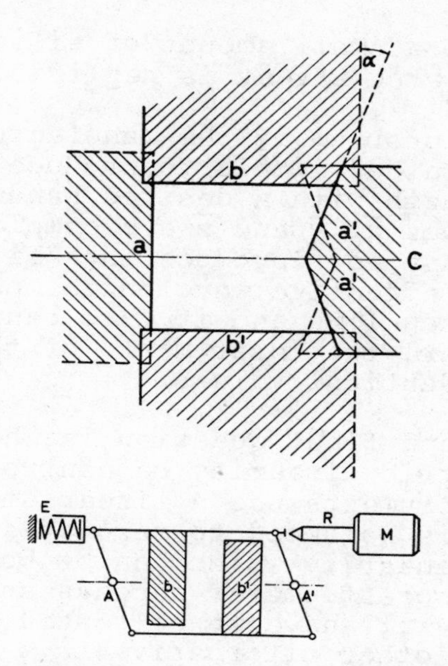

Fig. 7. Mechanical diaphragm with corrector.

measured with both geometries the readings usually do not coincide for the reasons already pointed out. Sphere geometry is coming more and more to the forefront and the recent proposal of ASTM includes this arrangement.

3. The Three Most Important Parts of the Filter Photometer

Three factors determine the accuracy of a filter photometer: the sphere, the filters and the measuring device.

It is evident that the sphere lining influences the illumination. Filters must be so designed that they incorporate a correction curve according to the applied light source. Tungsten illumination is commonly used and its emission curve has to be corrected to that of average daylight. Correction for the sensitivity of the applied photodetector must also be incorporated. The sphere is considered to be equally reflectant at all wavelengths. If yellowing or bleaching occurs the reflectance changes differently at different wavelengths and with it the illumination of the sample. The most stable sphere lining at present is magnesium oxide bound with a small amount of silicone (not silicate). The yellowing effect is negligible over years.

The second problem is the manufacturing of the filters. It is rather simple to produce filters for only one wavelength with a desired bandwidth but the tristimulus values $\bar{x}$, $\bar{y}$ and $\bar{z}$ are complex functions which are not easy to reproduce in a filter with fine tolerances. The effective wavelength must be kept within better than 1 mμ and if this cannot be achieved by a manufacturer, two instruments of the same brand will not give identical values.

The third and final question is, how accurately can the photometric linearity be controlled. Most present day instruments use a linear amplifier which, in some cases, is balanced to zero by a bias voltage. The degree of linearity which can be built into an amplifier or a potentiometer depends solely on price. In any case they all have to be tested and corrected very often. The other alternative is a mechanical diaphragm which allows an exact, linear change of light flux. An example of such a diaphragm is shown in Fig. 7. The sections b and b' are shifted by a parallelogram. One of the fixed sides of the square diaphragm can be folded so that the angle is variable. This provides a correction so that the lightflux can be changed linearly to better than 0,1%. The readings are made on the drum M.

Null balance instruments are usually more accurate than direct reading systems because they are less sensitive to voltage fluctuations.

The following pictures 8, 9 and 10 show some of the instruments designed according to the above mentioned principles.

4. Measuring of very small samples

The sphere arrangement makes it also possible to measure the reflectance of very small samples. If the image of a prism or grating is reduced by an achromatic objective it is easy to illuminate samples of a size of approximately 1 x 1 mm. Fig. 11.
The same objective can be used to project the slit of a monochromator 1 : 1 on a sample. Thus it is possible with a spectrophotometer of good quality to

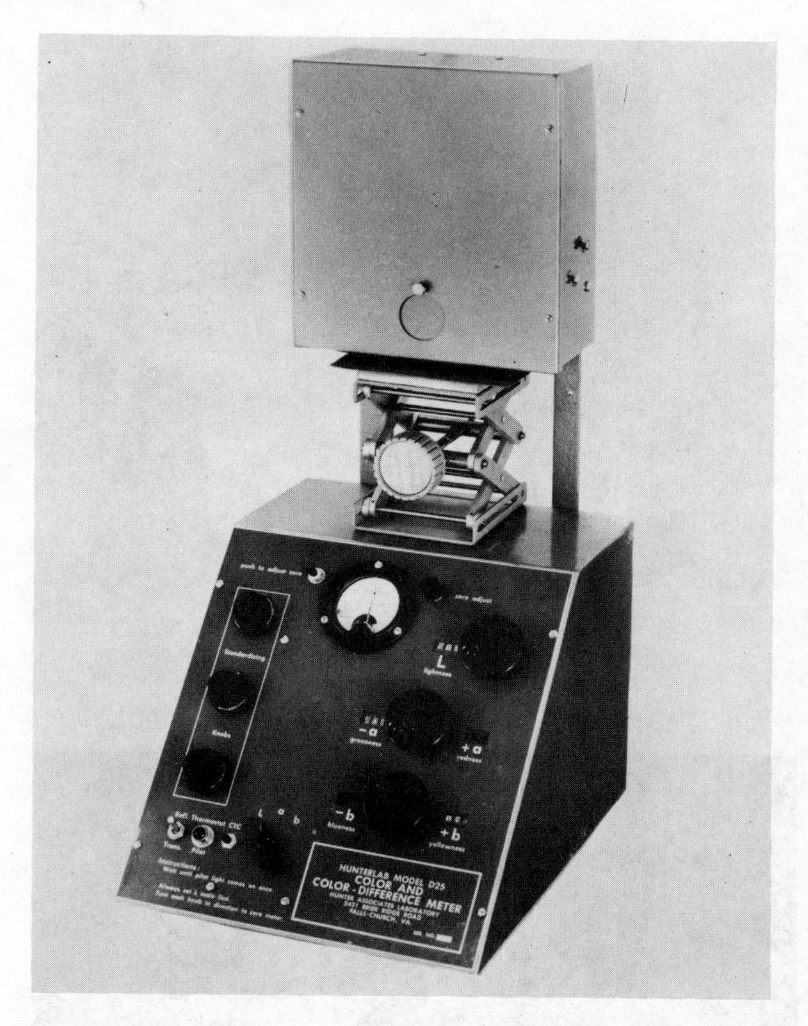

Fig. 8. Hunter Colorimeter (Hunter Assoc.)

measure the reflectance of a sample which has only 10 micron width. The height however has to be 3 to 5 mm. Samples of this kind are fibers, hairs or threads.

5. Measuring of Fluorescent Samples

In recent years many products on the market contain fluorescent additives. The fluorescence is mainly excited by ultraviolet light but the emission occurs in the visible range. For an absolute measurement of

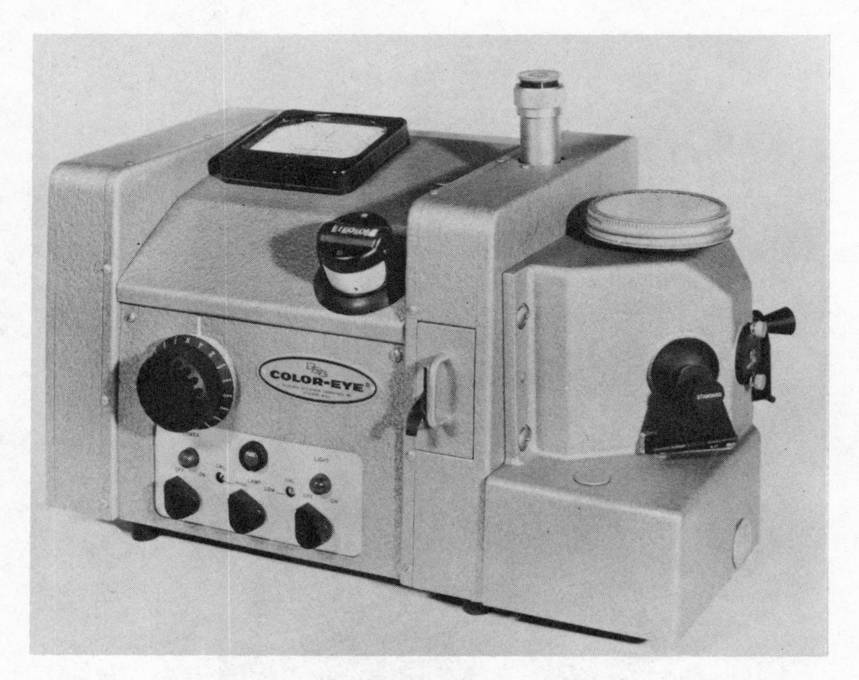

Fig. 9. Color-Eye Colorimeter (IDL)

Fig. 10. Elrepho (Carl Zeiss)

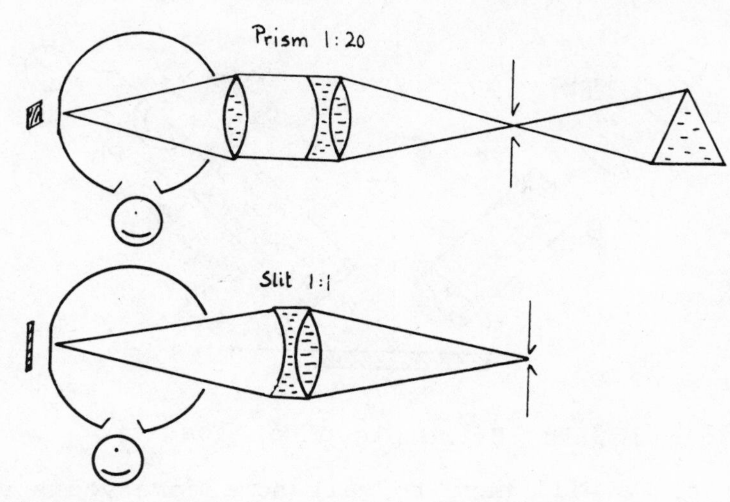

Fig. 11. Measuring principle of a very small sample with spectrophotometer and sphere attachment.

reflectance with the exclusion of the fluorescence two monochromators have to be applied with the sample in between. This type of instrument is not on the market as yet. Only filter photometers are available where the measurement is made with the full light of a xenon source. The xenon source emits a light which is very close to daylight and has enough ultraviolet to excite the fluorescence. If then a filter is inserted which cuts the ultraviolet off, two values can be read from the instrument: one with fluorescence and one without. Fig. 12. This will give relative values of the amount of the fluorescent additive.

Unfortunately the measurement of fluorescence is not standardized as is the case with reflectance mea-

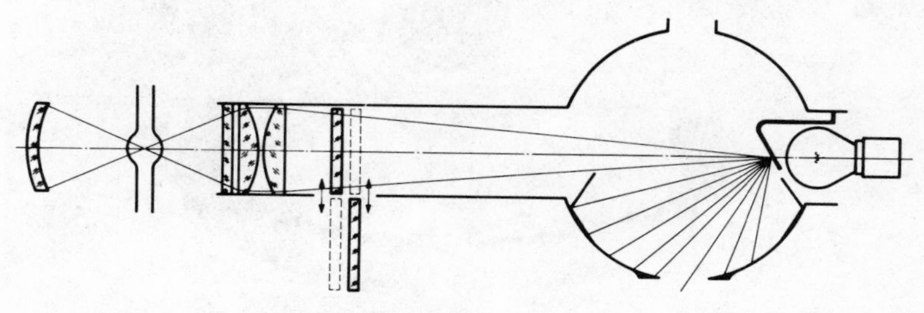

Fig. 12. Measuring arrangement to determine fluorescence with a filter colorimeter.

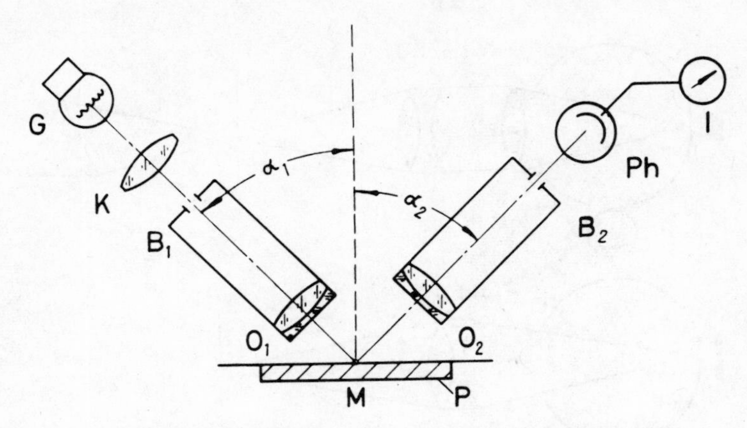

Fig. 13. Principle of a glossmeter

surement. We will have to wait some more years until the international committees have worked out this complicated procedure.

6. Specular Reflectance

The measurement of reflectance previously described involves only diffuse reflectance. Specular

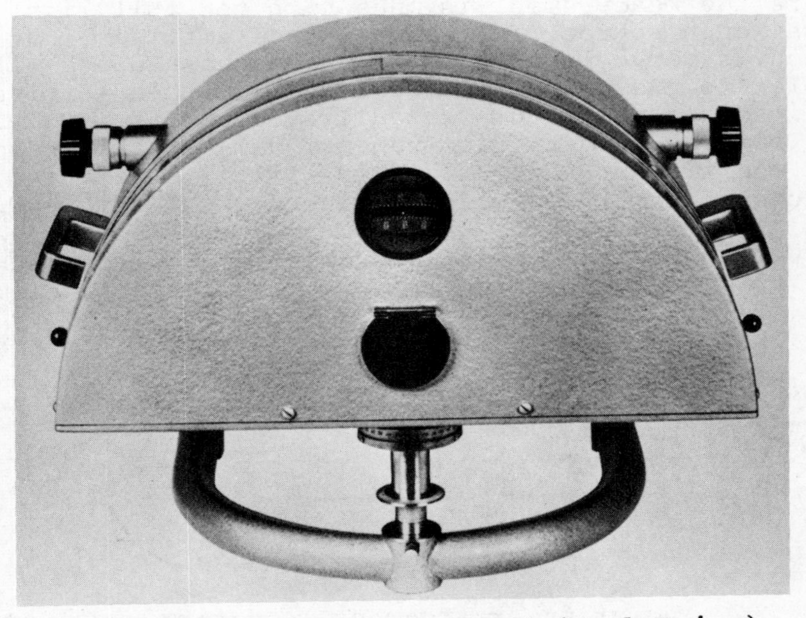

Fig. 14. The Glossmeter GP 2 (Carl Zeiss)

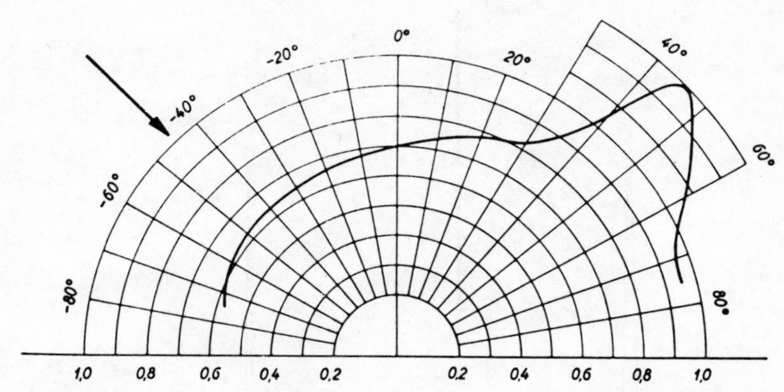

Fig. 15. Specular reflectance of cotton fabric

reflectance or gloss has to be eliminated by light traps, diaphragms or screens. If specular reflectance or gloss has to be determined, another kind of instrument must be applied.

No standard method exists at this time for measuring gloss. Thus values obtained from a glossmeter depend exclusively upon the design of the instrument used. The geometric quantities such as the geometry of illumination, the illumination angle, the aperture and the edge definition are dominant factors in the design of a glossmeter. Certain other parameters such as size of sample, spectral distribution and the selected standard influence the readings as well. The principle of a glossmeter is shown in Fig. 13. The illuminating collimator BK and the measuring collimator MK may be pivoted around an axis M in the surface of the sample P. The lamp L illuminates stop B1. Stop B2 determines the aperture of the observation beam. The optical systems O 1 and O 2 are completely identical.

Fig. 14 shows the outer design of such an instrument. The readings can be evaluated in two different ways: in polar coordinates or in rectangular coordinates.

Fig. 15 shows the specular reflectance of a cotton fabric relative to a reflection specimen. The arrow indicates the direction of the incident light. The curve obtained is nothing but a cross section of the spatial distribution of the reflected light. More common is the evaluation of the readings

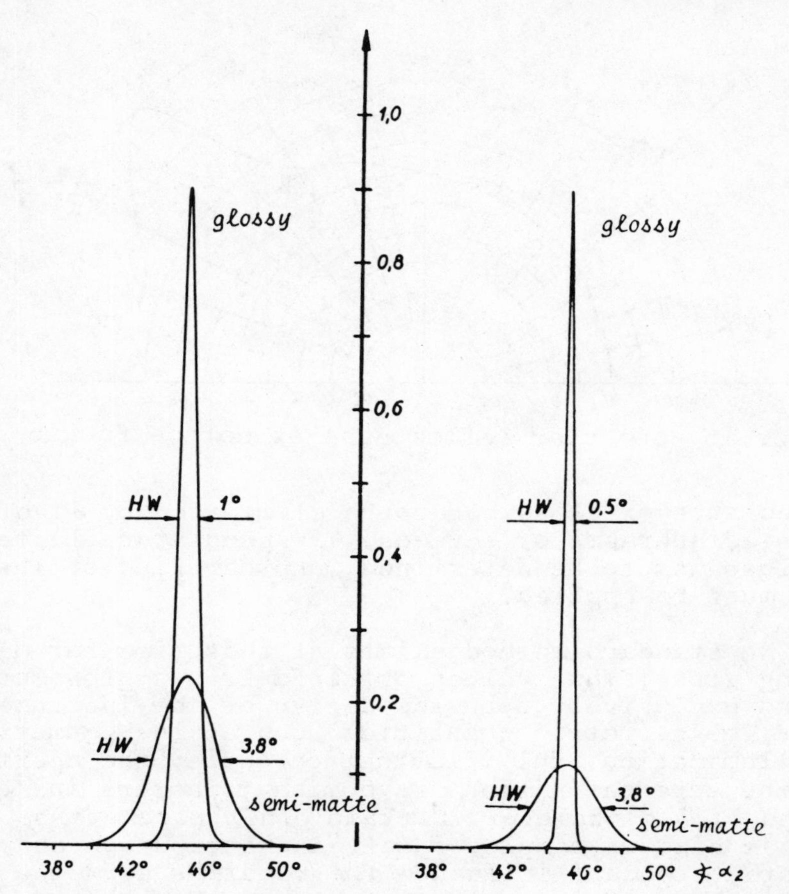

Fig. 16. Specular reflectance of photographic paper
The importance of the apertures

by means of a rectangular coordinate system. Fig. 16 shows the angular distribution of light reflected by photographic paper. The illumination angle is 45°. The observation angle is varied and the relative readings are plotted vs. observation angle. The reference standard was polished black glass.
In Fig. 16a the aperture of both collimators was set on 1° and in Fig. 16 b on 0.5°. The differences show clearly that the resolving power of such an instrument depends very much on the aperture of both collimators.

Looking through the ASTM specifications for the determination of gloss for different materials it is obvious that some of the specifications are made regarding the instruments available on the market at a time when there was a need for such a specification. We can only hope that the numerous committees working on a standardization will find a solution, so that instrument manufacturers can meet the requirements of the users better than they could in the past. What has been achieved for the measurement of diffuse reflectance should also be possible for the measurement of specular reflectance.

SOME NEW DIFFUSE AND SPECULAR REFLECTANCE ACCESSORIES FOR THE CARY MODELS 14 and 15 SPECTROPHOTOMETER

S. Hedelman and W. N. Mitchell

Cary Instruments, Monrovia, California

Diffuse Reflectance and Transmittance Spectrophotometry is an important analytical tool in the identification of materials and in the understanding of the structure of solids, crystals, and solutions.

Surface properties of materials affecting color and texture which are important in the controlled production of consumer goods have been studied using reflectance measuring techniques.

Significant impetus to the development of new instruments and accessories is partly due to special needs of space scientists, to characterize the surface of planets. Additionally, the intensive study of refractory paints and coatings in the solutions of the thermal balance equations is a prerequisite to successful scientific experiments in space vehicles.

A new accessory to the CARY Model 14 Spectrophotometer is now available, which has been designed to solve some of the instrumentation problems faced by chemists, physicists, biologists, and engineers.

The accessory, mounted to the right of the standard sample compartment of a Model 14 Spectrophotometer, is shown in Figure 1. The accessory is shown with the protractor and operating controls of the angular reflectance attachment installed. Access to the sample compartment of the accessory is from the right end. An important feature of the design is the ability to use the standard sample compartment for conventional measurement of specular transmittance or absorbance of solids and liquids without removing the attachment. This will be described in more detail later.

Accessory Mounted On CARY 14 Spectrophotometer

FIGURE 1

DESCRIPTION

The Cary Reflectance Accessory is intended for the measurement of reflectance and total diffuse transmittance using a two-inch photomultiplier tube and a 25 cm diameter integrating sphere. Provisions have been made for inclusion and exclusion of the specular component reflected from the sample. The sample has been placed in a horizontal plane to facilitate measurements on uncovered liquids, paste, and powdered samples as well as solid samples. Particular attention to sample handling problems facilitate the use of the accessory.

When recording spectra, the entire accessory is light tight, eliminating the need for excluding room light by improvisations.

Accessory Interior Showing 9-Inch Diffusing Sphere

FIGURE 2

When loading samples, arms which contain the sample and reference platforms are swung out of the sample compartment in a horizontal plane. This allows complete freedom for the operator in handling liquid or powdered samples with a minimum of spillage.

The large 25 cm diameter sphere improves uniformity of flux within the sphere. Large sample size reduces the effects of sample non-uniformity, which will be important with fabrics, powders, and other textured samples.

A series of attachments round out the utility of the Reflectometer. One attachment allows the measurement of total diffuse reflectance as a function of incident angle of solid samples located within the sphere. Another attachment permits the use of a PbS detector. By use of an additional attachment, the detectors may be moved from their normal position between the sample and reference material at the bottom of the sphere to the top of the sphere. In this location they are useful in the measurement of diffuse transmittance, specular reflectance, and in making normal absorption measurements of materials within the standard cell compartment, without taking losses normally associated with the use of integrating spheres. The "V" blocks at the top of the sphere allow the measurement of diffuse transmittance from sample cells placed at the entrance port to the integrating sphere.

For maximum flexibility, photomultiplier tubes with S-13, S-10, or S-20 spectral response characteristics are available. Unfortunately, no single phototube's spectral response curve is ideal for all forms of reflectance work.

The S-13, Dumont 7664, has a fused silica window which makes it useful from the short wavelength limit of the accessory out to about 6500Å. The S-20, RCA 7326, has a lime glass window which limits its short wavelength performance to about 3000Å. The long wavelength limit is near 9000Å. The tube has a higher signal-to-noise ratio when compared to the S-10 tube. Because of its cost, it should be used only where resolution or extended wavelength range justifies the S-20 surface. The S-10, RCA 6217, has a useful wavelength range of 2500-8000Å. In general this tube, the least expensive of the three, is adequate for tristimulus measurements in the color field and many other types of reflectance work.

All of the photomultiplier tubes normally furnished with the accessory are ten-stage devices. It has been our observation that when the commonly used dynode materials and structures are used in the manufacture of photomultipliers, no gain in signal-to-noise ratio can be obtained by an increase in the number of stages in the dynode structure. Also, we have observed that dynode voltages below the maximum available from the Model 14 power supply will give the best signal-to-noise ratio. Higher voltages tend to cause ionization of residual gases within the photomultiplier tube. The ions create large noise spikes which more than obscure the effects of additional gain in amplification due to the higher voltages.

For work at wavelengths longer than about 7000Å, the lead sulfide detector exhibits a higher signal-to-noise ratio than any uncooled known photomultiplier tube. Even the cooled tubes exhibit lower signal-to-noise ratio than the lead sulfide detector beyond 10,000Å. For work at long wavelengths, a lead sulfide detector modification is available.

THEORY

Integrating sphere theory has been discussed by any number of authors. They have used many different mathematical models and methods to attack the problem. One of the more eloquent and complete discussions is due to Jacquez and Kuppenheim[1].

OPTICAL SYSTEM

The optical system for diffuse reflectance and for tristimulus measurement is shown in Figure 3. The system between lamp A or c and the sample space T is unchanged from that of the CARY 14.

In the Model 14 signal generator compartment, light is chopped at 30 cps and alternately directed along sample and reference paths T and T′ as shown in Figure 3.

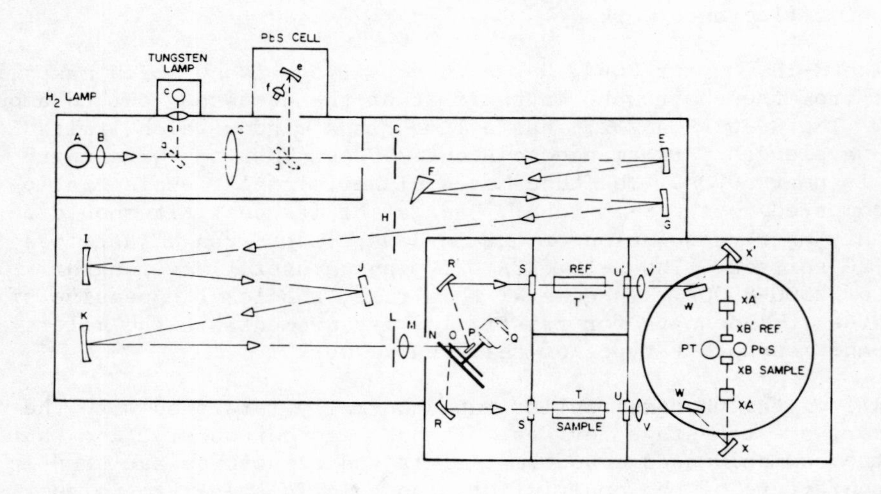

OPTICAL SYSTEM
CARY MODEL 14 SPECTROPHOTOMETER
WITH 50-400-000

FIGURE 3

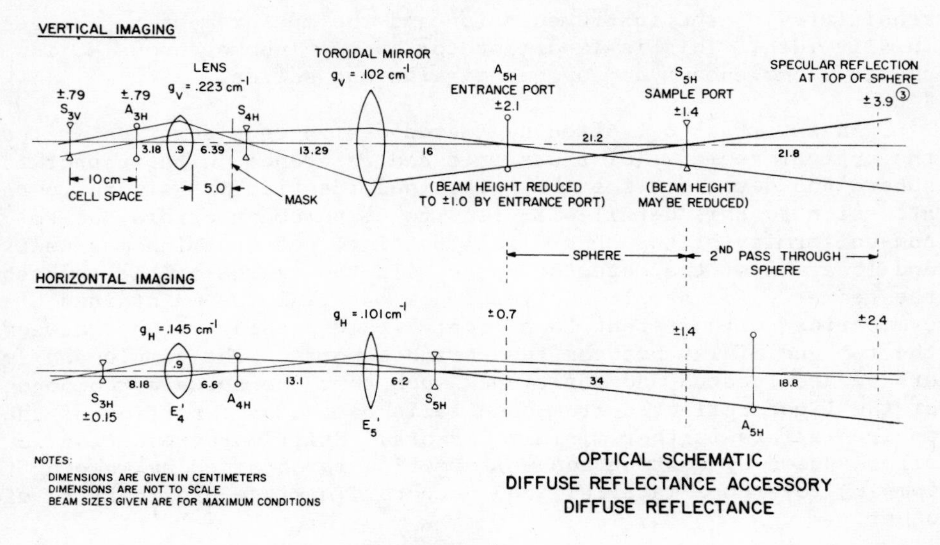

FIGURE 4

The instrument records the ratio of sample to reference signals on a linear or logarithmic scale as a linear function of wavelength.

The light passes from the standard Model 14 cell compartment into the accessory, passing first through the lens V to the flat mirror W, where it is directed towards the toroidal mirror X, which redirects the light towards the flat mirror XA. The flat mirror XA now directs the light downward through the sphere entrance port to the sample XB at the bottom of the sphere. The light reflected by the sample is multiply reflected within the sphere until it is lost or it reaches the detector.

An optical system for reflectance work using an integrating sphere should be designed to minimize the total port area, since the efficiency of the sphere is inversely proportional to the sum of the port areas and the equivalent dark area of the sphere wall. Minimum port area for the system is achieved when equal images of the vertical slit and aperture are located at the entrance port and the sample port of the sphere. Figure 4 is a schematic of the optical system. The detector location is not shown in the optical schematic. It is located near the sample and in a position so that radiation reflected directly from the sample must strike the sphere wall before hitting the detector. Should it be located otherwise, the measurement would contain an error dependant upon the gloss of the sample.

Intentional placement of an aperture image on the sample maintains a constant beam size on the sample, thus reducing the sensitivity of the instrumentation and the measurement to changes in slitwidth. This is important for opaque inhomogeneous solids and for homogeneous non-opaque liquids and solids.

An important but often neglected design consideration is that the optical geometry of the sample and reference, with respect to sphere and detector, must be maintained identical. Failure to pay attention to this detail will lead to measurement errors due to non-uniformity of the sphere wall coating, detector inhomogeneity, and location of the detector ports. In the design of the accessory, the geometry for sample and reference material was maintained symmetrical with respect to a great circle passing midway between the two and midway between the entrance ports. The sample and reference are located and positioned such that the specular component of the light reflected from each falls on the same portion of the sphere wall to further minimize errors. Still more reduction in error caused by coating non-uniformities is obtained by keeping the sample, reference material, and detector in close proximity to each other.

SAMPLE HANDLING

As stated before, when making diffuse reflectance measurements or total reflectance measurements, the detector must avoid being hit by radiation reflected directly from the sample. This is one of the basic assumptions of integrating sphere theory. The attachment has three optional detector positions that serve the purpose of avoiding direct reflection from the sample. It is necessary that the user select the proper detector position for the type of measurement to be made.

When diffuse reflectance or tristimulus measurement are to be made, the detector is mounted in its lower position. The sample is mounted on the sample holding arm and a magnesium oxide smoked cup is mounted on the reference holding arm.

Elimination of the specular component may be accomplished by using the above arrangement if the operator removes the spherical cap, located at the top of the sphere.

For transmittance measurements of clear solutions such as dyes, the sample is placed into the standard Model 14 cell compartment. The detector should be placed at the top of the integrating sphere. Mirrors are placed into the cups that normally hold the sample for diffuse reflectance work. When measurements are made in this manner, the light is reflected directly onto the detector without making use of the integrating properties of the sphere. This avoids energy losses which would result from multiple reflections within the sphere.

To make measurements of diffuse transmittance, the sample is placed into the holder "V" block at the entrance port of the sphere (Figure 2), and the detector at the top of the sphere. The sample and reference cups are coated with magnesium oxide smoke.

Specular reflectance measurements may be made with the detector in the upper position. The sample is placed on the sample holding arm. A reference standard of specular reflectance is placed on the reference holding arm. This method tends to ignore the light diffusely reflected from the sample.

DRY GAS FITTINGS

The accessory is equipped with separate gas fittings for the upper sphere chamber and the lower sample chamber. This feature serves the purpose of reducing purge time when introducing new samples and by maintaining a slight positive pressure of clean gas in the sphere chamber, thereby reducing the influx of dust and hydrocarbons into the sphere. This reduces the rate of deterioration of the sphere coating, especially in localities where air pollution is great.

UTILITY ACCESS

A removable plate is provided at the rear of the accessory to accept electrical connectors, plumbing, and other customer modification as may be required for unique sample handling problems.

SPHERE ACCESS

The top cover and the end cover plate of the accessory are held into place with Dzus fasteners which allow the operator to quickly bare the interior of the accessory while installing attachments.

SPHERE EFFICIENCY

Using equations derived by Jacquez and Kuppenheim[1], the efficiency of the integrating sphere with photomultiplier tube is 21% and with lead sulfide detector is 12%.

The sphere is coated by depositing smoke from magnesium chips of high purity.

For tristimulus work, the sphere may be painted with Eastman Kodak Co. #125 High Reflectivity White Paint. The paint is highly stable, and over the wavelength range of 4000-21,000Å, nearly as

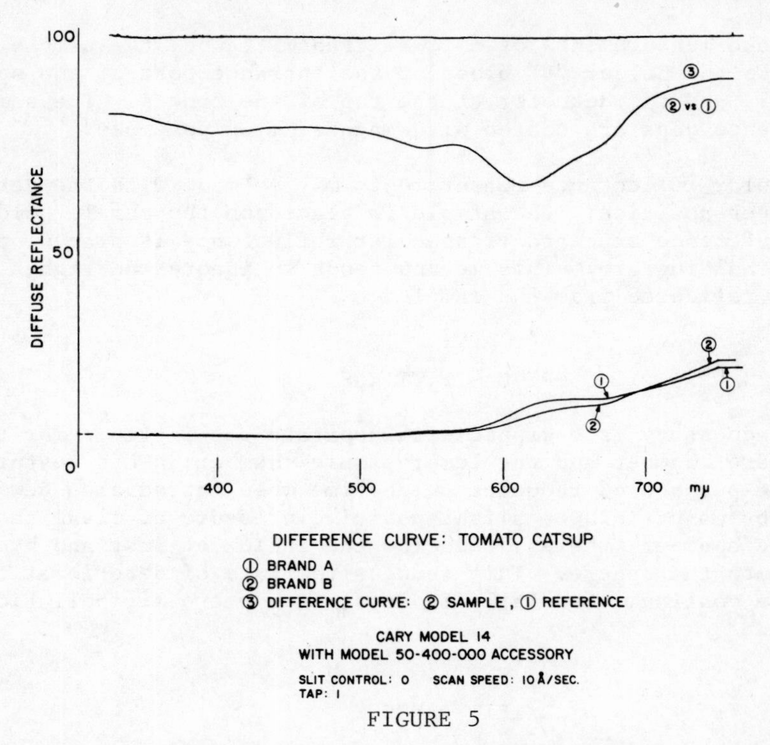

FIGURE 5

reflective as MgO smoke. By using maximum gain and pen period control accessory 1490800, the wavelength range can be extended to 25,000Å. At wavelengths shorter than 3500Å, the reflectivity falls quickly.

The CARY Model 14 Reflectometer Accessory has been in use for approximately six months. While its full capability has not been explored, the attention to design aspects (which have been obstacles in the measurement of diffuse reflectance and transmittance) promises a versatile and accurate instrument for the laboratory. These features are:

Large constant beam size for inhomogeneous samples.

Symmetry of sample and reference location with respect to entrance and exit ports.

Horizontal sample handling system for measuring powders and uncovered liquids.

Wide wavelength range with continuous scan from 2500-25,000Å.

Flexibility to allow various measurements with a minimum of interchanges in hardware.

Figure 5 shows spectra of typical tomato catsup samples in the region where color is an important quality. The enhancement of color differences is most apparent when difference measurements are made. Figure 6 shows the effects of water adsorption on indicating Drierite.

Figure 7 shows the efficiency of the instrument in terms of the resolution in the near IR of the complexed spectra of Holmium Oxide. Resolution of better than 5Å in the visible has also been obtained.

A new microspecular reflectance accessory for the Models 14 and 15 is also now available.

The accessory mounts inside the Model 14 or 15 standard sample compartment and is interposed in the sample beam.

Beam size is 1.5 x 1.5 mm in the Model 14, using 1/3 slit heights, and is 3.0 x 1.5 mm in the Model 15 using full slit heights.

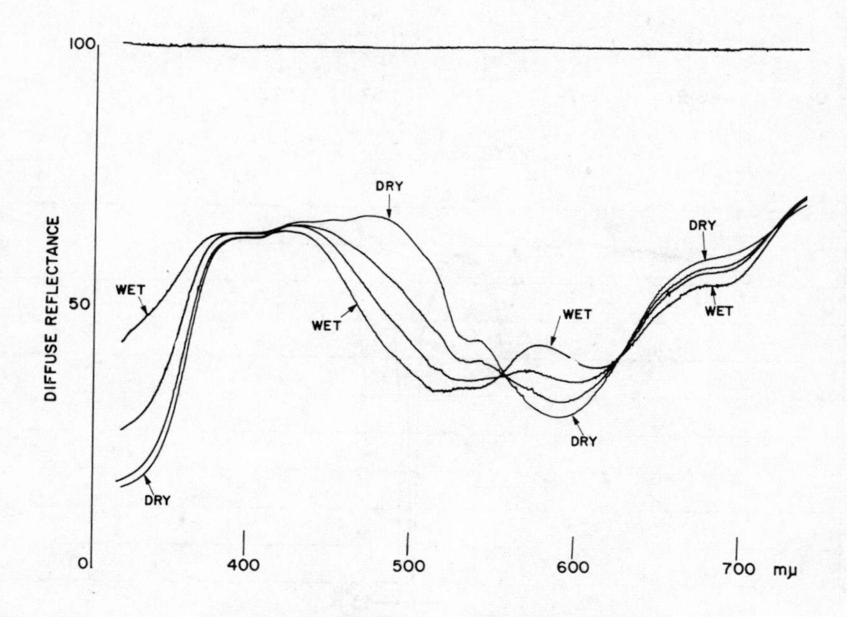

SAMPLE: GRANULAR INDICATING DRIERITE

CARY MODEL 14
WITH MODEL 50-400-000 ACCESSORY
DYNODE TAP: I SENSITIVITY: O

FIGURE 6

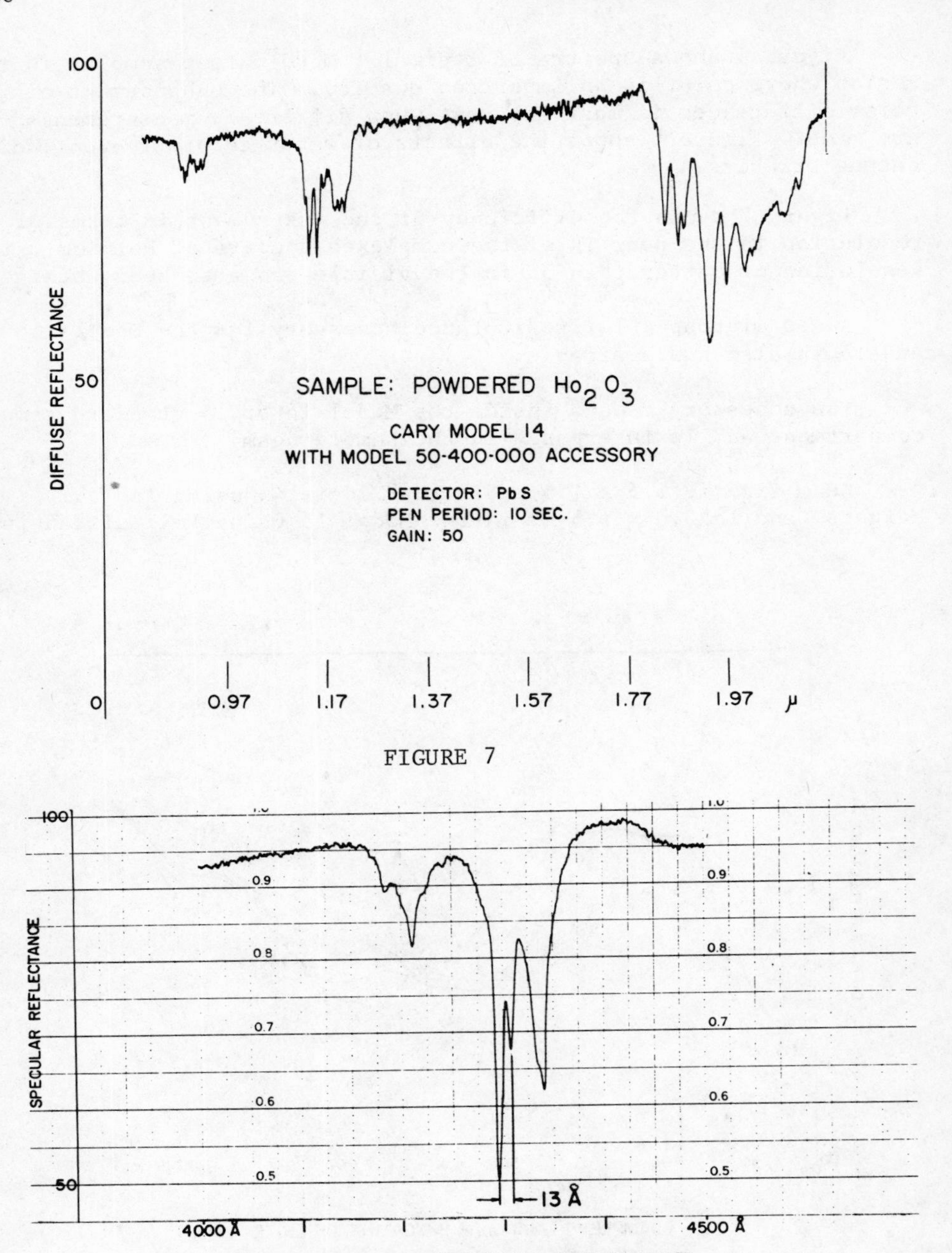

FIGURE 7

SAMPLE: NEODYMIUM CHLORIDE
CRYSTAL (4 x 3 x 2 mm)
CARY MODEL 14
WITH MACROSPECULAR REFLECTANCE ACCESSORY

FIGURE 8

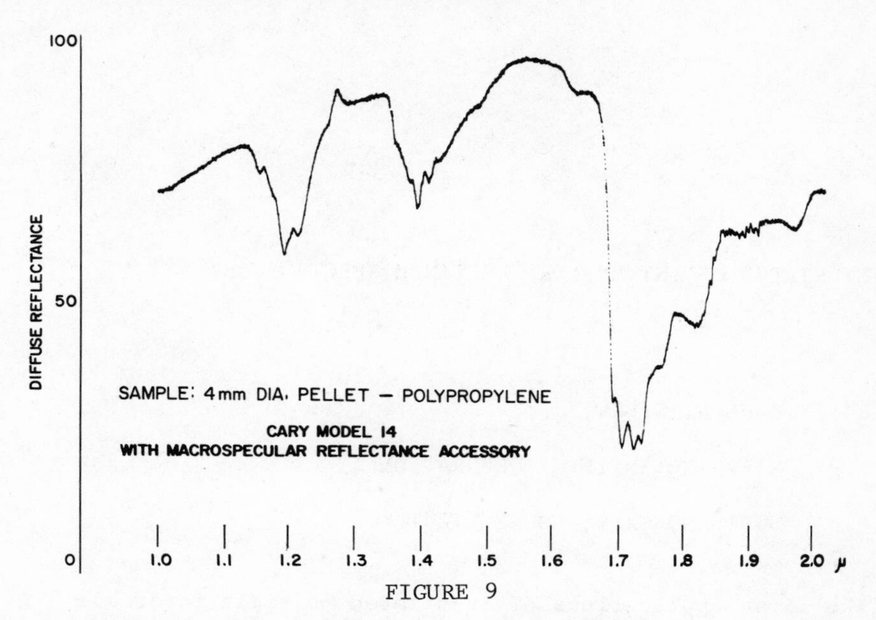

FIGURE 9

In the simplest version, the sample is placed on top of a platform containing a small aperture. This sample handling system is suitable for solids.

For liquids or powders, a more complicated version has been designed in which the sample platform is below the beam center line. The sample is loaded in a cuvette and the height is adjusted until the surface is positioned at the correct height.

Reflection from a toroidal mirror turns the beam towards the sample at an incident angle of 13°. The efficiency of the system is approximately 60% when measuring a specular sample.

Using a mirror of known specular reflectance, absolute values may be obtained for the sample. Diffusely reflecting samples may also be used.

The efficiency of the system when measuring a good diffusing material, such as MgO, is 4%.

Figure 8 shows the spectrum of Neodymium Chloride crystal 4 x 3 x 2 mm with a spectral bandwidth of 1.5Å

Figure 9 shows the spectrum of a 4 mm diameter Polypropylene pellet. Despite the simplicity of this accessory, it should prove useful in the study and identification of crystals, small samples, and in the exploration of surface homogeneity.

1. John A. Jacquez and Hans F. Kuppenheim, J. Opt. Soc. Am. 45:460 (1955)

CURRENT STATUS OF INTERNAL REFLECTANCE SPECTROSCOPY

PAUL A. WILKS, Jr. •

WILKS SCIENTIFIC CORPORATION

SOUTH NORWALK, CONNECTICUT

The first applications of frustrated multiple internal reflectance or attenuated total reflectance to infrared spectroscopy were reported by Harrick (1) and Fahrenfort (2) in 1959. Because their work was done independently, the two different terminologies have been used in the literature and a certain amount of confusion has resulted. It should be understood that both FMIR and ATR refer to the same phenomenon; the more general term "Internal Reflection Spectroscopy" will be used in this discussion.

The internal reflection technique is extremely important to spectroscopists for it provides a method for obtaining infrared spectra directly on solid materials and surfaces with little or no sample preparation. A wide variety of materials such as films, powders, plastic samples, adhesives, fabrics and fibers, foams, coated metals and paints and finishes, are amenable to analysis by internal reflection. When the proper operating parameters are set up, it is possible to obtain spectra by internal reflection that are so nearly similar to transmission spectra that existing spectra libraries may be used for qualitative identification.

An excellent mathematical treatment of the internal reflection phenomenon will be found in "Internal Reflection Spectroscopy" by Harrick (3). The following geometrical description, excerpted from (4), though not as rigorously correct will provide a basic understanding:

"When a beam of radiation enters a prism it will be reflected internally when the angle of incidence at the interface between sample and prism is greater than the critical angle (a function

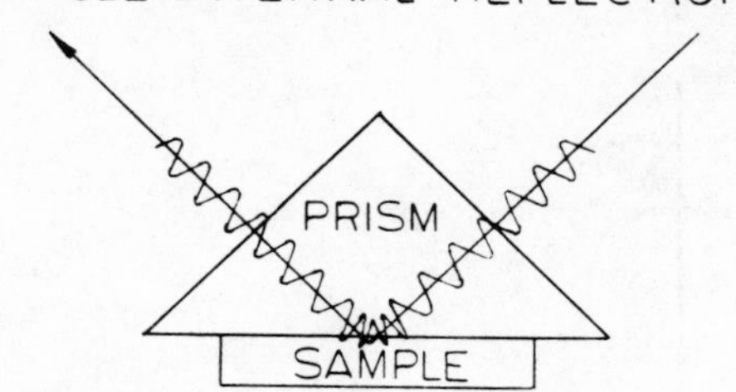

Figure 1 Internal Reflection Effect

of refractive index. On internal reflection all of the energy is reflected. However, the beam appears to penetrate slightly beyond the reflecting surface, and then return, as shown in figure 1.

"When a material which selectively absorbs radiation is placed in contact with the reflecting surface, the beam will lose energy at those wavelengths where the material absorbs owing to an interaction with the penetrating beam. This attenuated radiation, when measured and plotted as a function of wavelength by a spectrophotometer, will give rise to an absorption spectrum characteristic of the material.

"The depth to which the radiation penetrates is a function of (1) the wavelength of light, (2) the refractive index of both the reflector and sample, and (3) the angle of incident radiation. The apparent depth of penetration ranges from a fraction of a wavelength up to several wavelengths. If the depth of penetration versus angle of incidence is plotted in a region where the sample absorbs (figure 2), the penetration will increase most rapidly when the angle of incidence at the interface between the sample and prism is very near the critical angle but it is relatively slight at angles well removed from the critical angle. Nearly all the energy goes into the sample when the critical angle is exceeded.

"Figure 3 contains a typical plot of the index of refraction of a material versus wavelength in the vicinity of an absorption band. Note that the index of refraction undergoes a radical variation in this region. If a reflector of index n_1 is selected, there is a point at which the index of the sample is greater than that of the reflector. At this wavelength, there is no angle of incidence at which internal reflection can take place and nearly all the energy will pass into the sample. The recorded internal reflection absorption band will thus be very strong but broadened toward long wavelength and hence greatly distorted as compared to that measured by transmission.

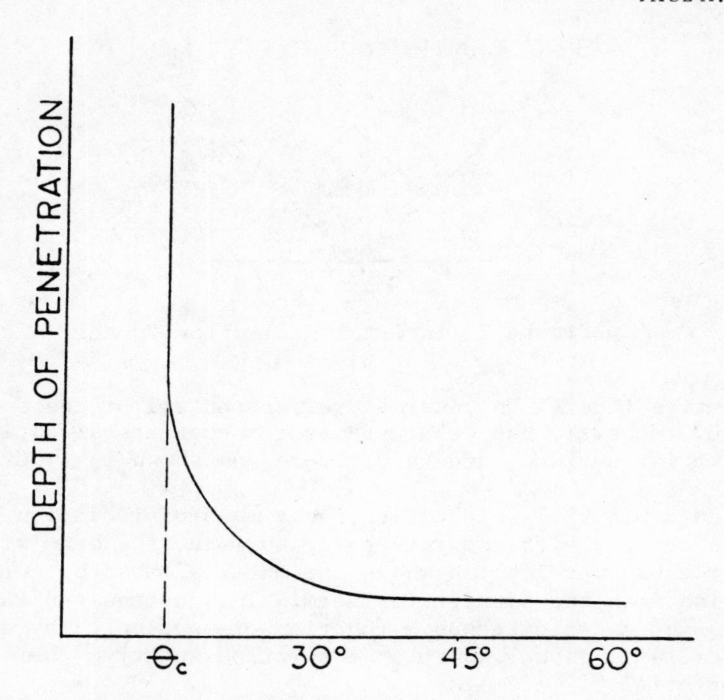

Figure 2 Depth of Penetration vs. Angle of Incidence

"When a reflector of index n_2 is chosen, there is no point at which the index of the sample exceeds it. There is a region on the long wavelength side of the band where the index of the sample does come close to that of the plate which is the same as though the angle of incidence was increased.

"Hence, although the resulting internal reflection absorption band is less distorted than in the first case, there is considerably more apparent absorption on the long wavelength side. Note also that the total absorption is reduced.

"In the third case, n_3 is considerably greater than that of the sample. Under this condition the index variation of the sample has practically no effect on the shape of the band although the total absorption in the band is small.

"Thus, in order to obtain internal reflection spectra that are nearly identical to transmission spectra, a relatively high index reflector should be used. Also an overall angle of incidence should be selected that is far enough from the average critical angle of the sample versus reflector so that the change of the critical angle through the region of changing index has a

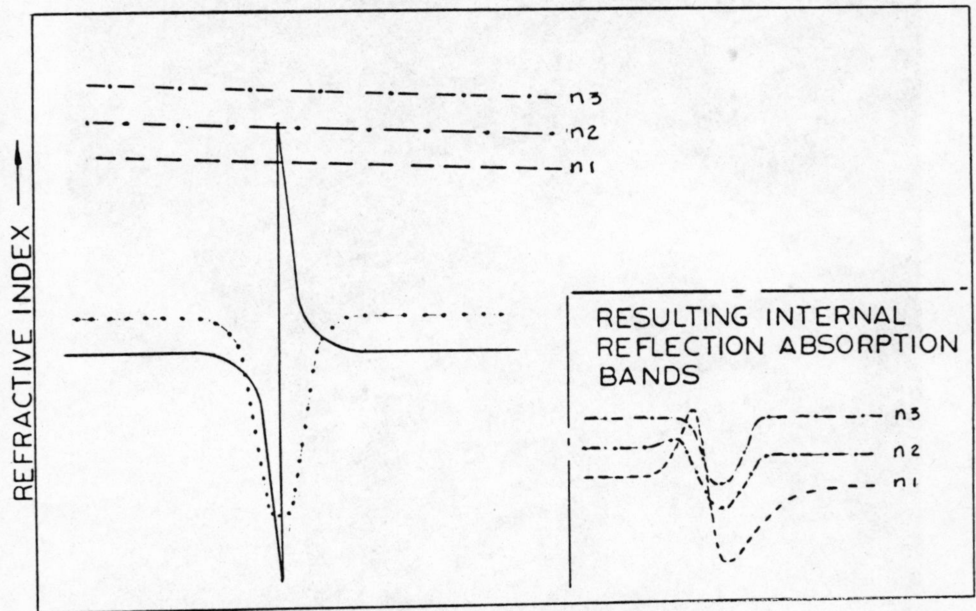

Figure 3 Refractive Index vs. Wavelength

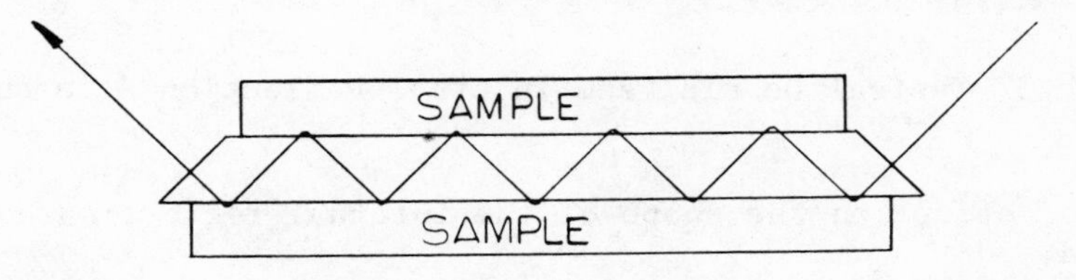

Figure 4 Multiple Internal Reflection Effect

Figure 5 Typical Double Beam Internal Reflection Attachment

minimum of effect on the shape of the internal reflection absorption band.

"The energy exchange in case three is relatively small so that the absorption band produced by a single reflection is weak. It is possible, however, to multiply the amount of absorption by multiplying the number of reflections as in figure 4. This is analogous to increasing the path length in transmission cells.

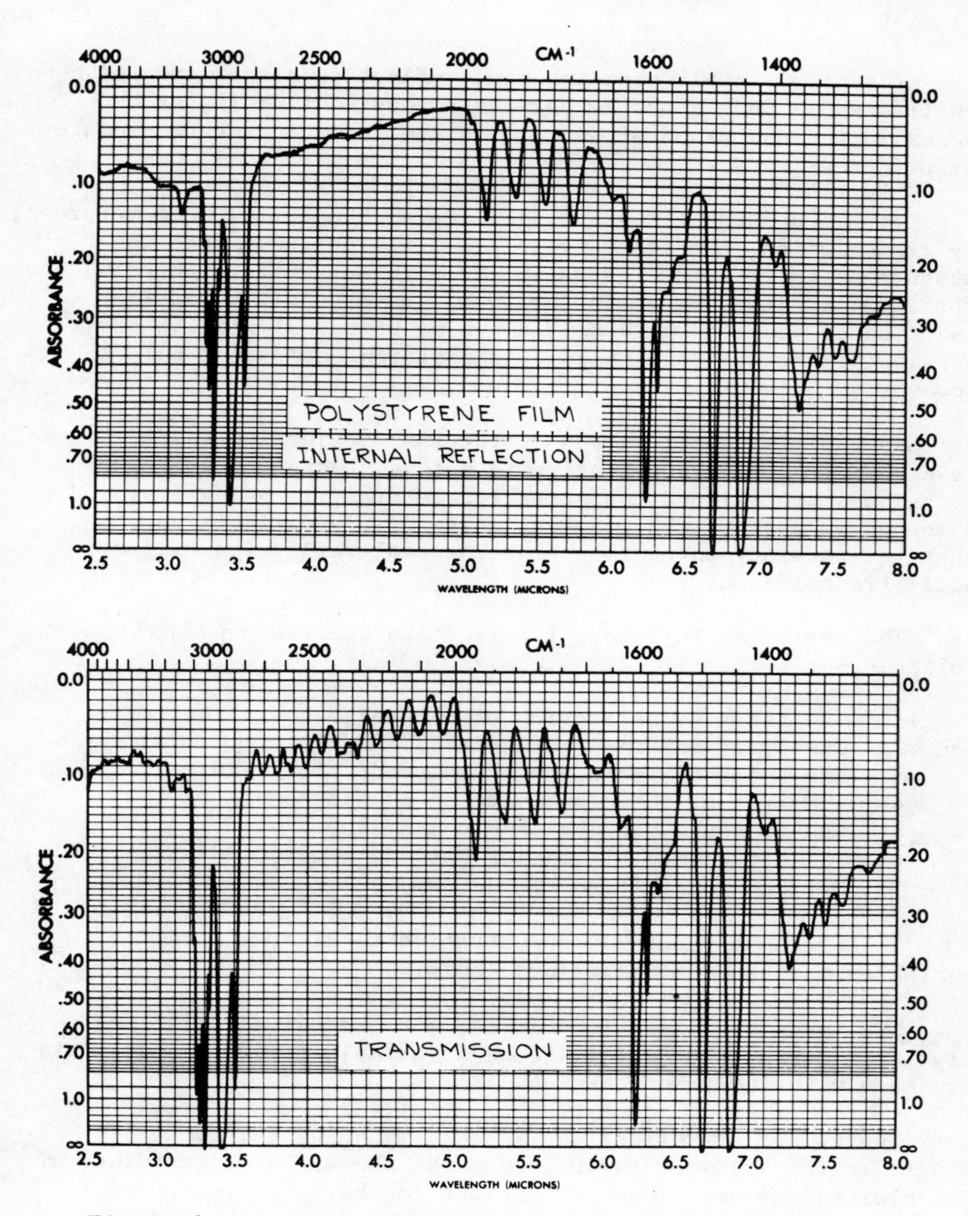

Figure 6 Transmission and Internal Reflection Spectra

The multiple reflection approach can thus produce undistorted spectra of any desired intensity provided there are enough reflections."

A typical double beam internal reflection attachment is shown in figure 5. Such a system permits accurate balancing of both beams in the spectrometer as well as the recording of compensated or differential spectra.

Figure 6 shows two spectra of the same material, one recorded by transmission, the other by internal reflection. Note the absence of interference bands in the internal reflection curve and the relative increase in intensity of the absorption bands as a function of wavelength in this curve as compared with the transmission spectrum. Band shapes and positions are identical, however.

The initial applications of internal reflection spectroscopy have been generally qualitative in nature. As equipment and techniques are refined, quantitative accuracy is being improved, although because of the many variables involved, it is doubtful that the accuracy obtainable from internal reflection will ever equal transmission.

The variables in internal reflection spectroscopy are the following:

1. The index of refraction of the reflector plate.
2. Number of internal reflections.
3. The angle of incidence of the radiation beam.
4. The area of the plate covered by the sample.
5. The pressure between sample and plate.
6. The efficiency of contact between sample and plate.
7. Surface composition of the sample.

To obtain reproducible quantitative results, all these variables must be controlled precisely.

The index of refraction varies from crystal material to crystal material, but ceases to be a variable if all measurements are made with the same material.

The number of reflections is determined by the ratio of length to thickness of the plate and by the angle of incidence of the radiation beam. These conditions can be fixed relatively easily.

The angle of incidence is set at 45° or 52° on many multiple internal reflectance attachments, adjustable to 30°, 45° or 60° on more recent models, and fully adjustable at any angle on the

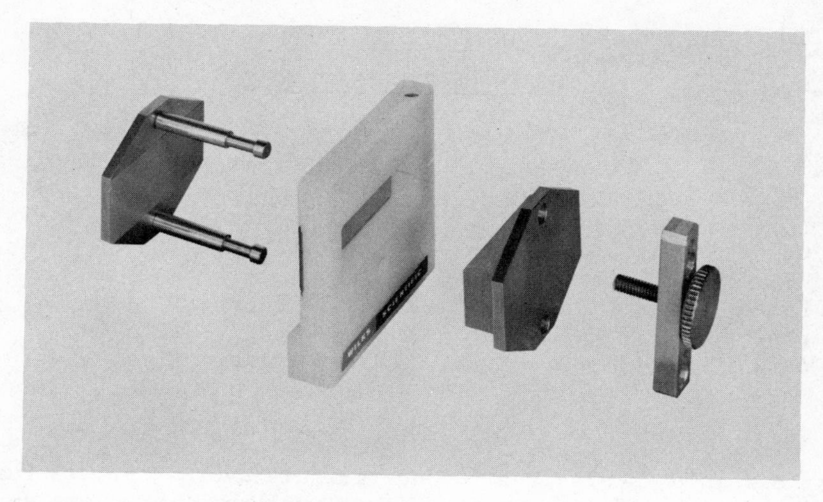

Figure 7 Fixed Plate Sample Holder

second generation internal reflection equipment now becoming available. (See below.)

The area of the sample in contact with the plate is somewhat analogous to path length in transmission spectroscopy. For quantitative accuracy, successive samples should be identical in size and if possible, should be contacted to the same portion of the plate.

The biggest source of error in internal reflection spectroscopy is in reproducing the exact contact conditions from sample to sample. This can be done reasonably well with smooth soft samples, but becomes increasingly difficult with rough or hard samples. Newly developed sample holders such as that shown in figure 7 make the clamping of the sample against the plate more precise. For best results, this mount should be coupled with an adjustable torque screw-driver or wrench to insure reproducible pressure settings.

Since internal reflection spectroscopy is essentially a surface measuring technique, the spectrum obtained through its use is representative of the top few microns only and may or may not be representative of the bulk sample.

When all variables are understood and carefully controlled, quantitative results reproducible to within $\pm 3\%$ should be attainable.

An application area of increasing interest for internal reflection is the study of surfaces. The penetration of the beam

TABLE I

DEPTH OF PENETRATION

($n_2 = 1.5$)

	30°	45°	60°
Ge	.091	.041	.002
KRS-5	i	.290	.113
AgCl	i	i	.316

into the sample is predictable and variable over a certain range. Hence, the technique can be used for a variety of problems such as the measurement of coating thickness, the determination of surface degradation, the detection of plasticizer bleeding and many others.

The depth of penetration can be calculated from the following equation (3):

$$(1) \qquad dp = \frac{\lambda_i}{2\pi(\sin^2\theta_i - n_{21}^2)^{1/2}}$$

where: n_{21} = refractive index of the sample divided by the refractive index of the crystal.

λ_i = λ/n_1 = wavelength of radiation in the crystal.

θ_i = the angle of incident radiation at the interface.

The effective path through the absorbing medium is determined by the following equations:

$$(2) \qquad \frac{de_\perp}{\lambda_1} = \frac{n_{21}\cos\theta}{\pi(1 - n_{21}^2)(\sin^2\theta - n_{21}^2)^{1/2}}$$

and (3)

$$\frac{de_\parallel}{\lambda_1} = \frac{n_{21}\cos\theta(2\sin^2\theta - n_{21}^2)}{\pi(1 - n_{21}^2)[(1 - n_{21}^2)\sin^2\theta - n_{21}^2](\sin^2\theta - n_{21}^2)^{1/2}}$$

TABLE II

BENZENE SAMPLE THICKNESS IN MICRONS
STANDARD ABSORBERS VS. E.P.L.

	$\tau(1960) = 10^2 A$	E.P.L. = Nd	N
Ge 1mm 45°	7.8	8.1	40
KRS-5 2mm 45°	28.0	29.0	20
AgCl 2mm 60°	17.5	18.0	11.5
	$\tau(850) = 10^3/2.2$ (A)	E.P.L. = Nd	N
KRS-5 2mm 45°	68.7	68.0	20
AgCl 2mm 60°	43.2	42.9	11.5

When only one reflection is used, the effective path length (EPL) equals de. In a multiple reflection plate:

$$\text{EPL} = N d_e \tag{4}$$

where: N = Number of reflections = $\frac{\text{Length}}{\text{Thickness}} \cdot \cot\theta_i$

It can be seen from (2) and (3) that d_e decreases as θ_i increases and as n_1 increases. Table I gives some representative depths of penetration for different materials and angles of incidence.

A new variable angle, multiple reflection attachment is shown in figure 8. With such an accessory it is possible to set θ_i to any desired angle accurately and reproducibly so that desired penetrations can be selected.

Table II shows a comparison of calculated EPL's vs. measured absorbances from the benzene spectra in figure 9.

The next figure (10) shows that the beam does indeed change penetration with angle. The sample is a polyethylene coated cellophane. Little or no cellophane absorption is shown in the top spectrum. In the lower curve, with a reduced angle of incidence, cellophane absorption is more apparent.

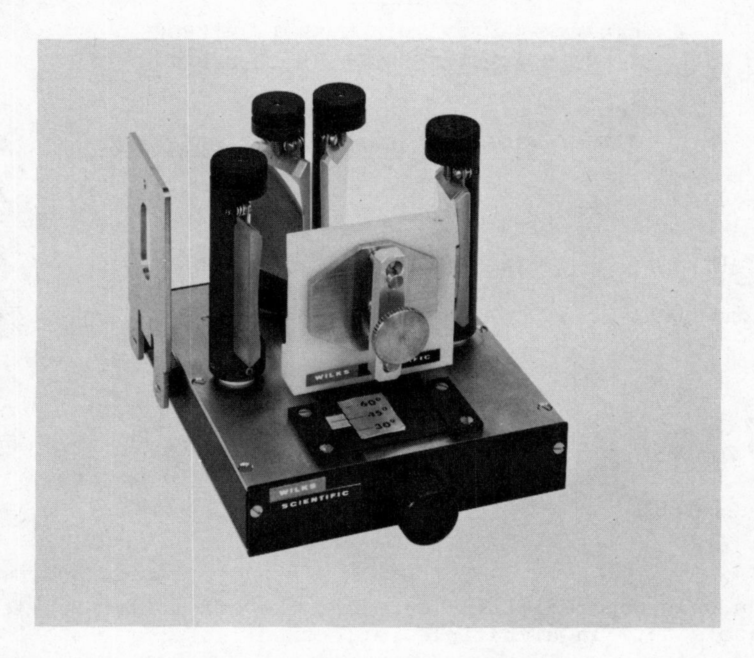

Figure 8 A Variable Angle Multiple Reflection Attachment

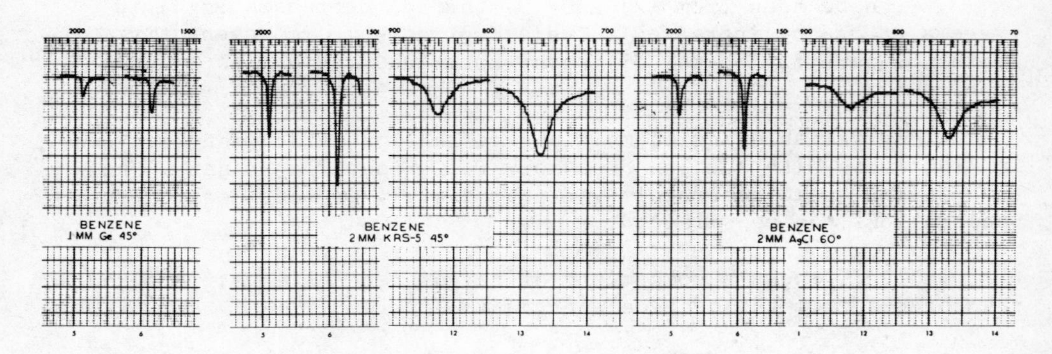

Figure 9 Benzene Spectra

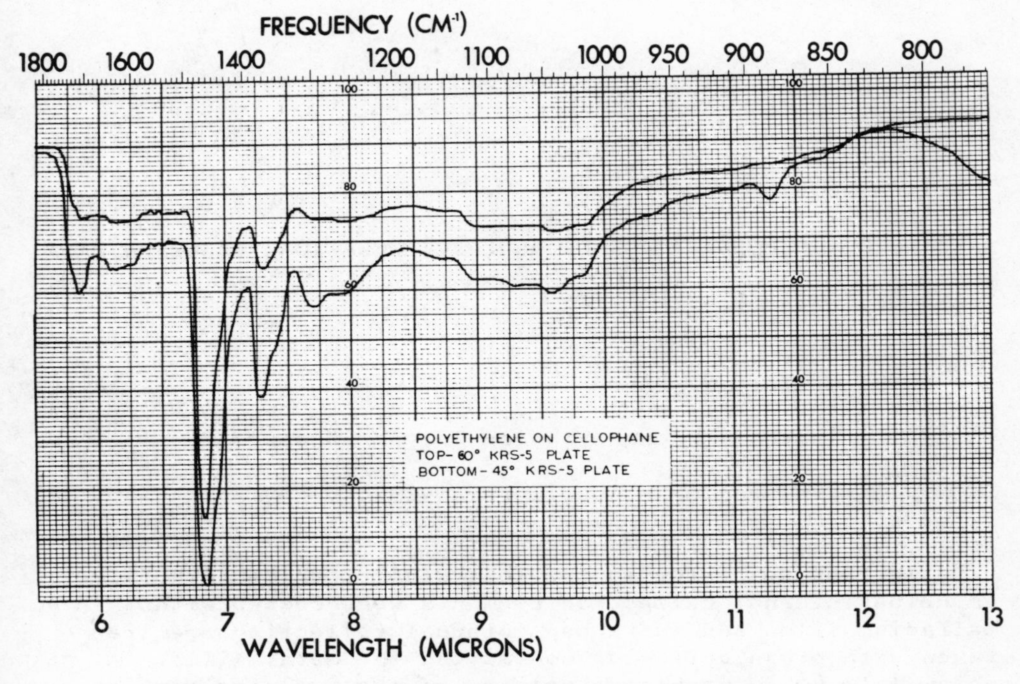

Figure 10 Polyethylene Coated Cellophane

In summary, the internal reflection technique, since its introduction eight years ago, has demonstrated its usefulness as a qualitative tool. As new and more precise equipment becomes available, it is now possible to use the method for accurate quantitative measurements approaching the precision of transmission spectroscopy. In addition, some of the many variables such as penetration may be used to advantage in special sampling problems.

REFERENCES

1. Harrick, Annals of New York Academy of Sciences, Vol. 101, Art. 3, Pages 928-959. 1963.

2. Fahrenfort, Spectrochemica Acta, Vol. 17, Pages 698-709. 1961.

3. Harrick, "Internal Reflection Spectroscopy," Interscience, New York. 1967.

4. "Internal Reflection Spectroscopy," Vol. 1. Wilks Scientific Corporation. 1965.

SPECTROSCOPIC OBSERVATION OF AN ELECTROCHEMICAL REACTION BY MEANS OF INTERNAL REFLECTION

Wilford N. Hansen

North American Aviation Science Center

Thousand Oaks, California 91360

In 1962 we performed the following experiments. Silver chloride internal reflection elements were coated with thin palladium films and infra red internal reflection spectra taken with organic liquids contacting the metal films. A metal film 40 Å thick gave spectra of the organics similar to attenuated total reflection (ATR) spectra. A thicker film (*ca.* 175 Å thick per chemical analysis) having a resistance of 40 Ω per square also gave spectra of the organics, but with less intensity. These simple experiments showed that spectra of the region close to the electrode could be taken *through* a thin metal electrode. These results were reported by Hansen and Osteryoung[1] with the suggestion that this type of optical set-up could be used to study electrochemical reactions *in situ*. It was also suggested that an electrical conductor like germanium could serve as both internal reflection element and electrode. Later in 1962 glass multiple reflection elements were coated with various metals, and spectra of dyes taken in the visible. When the metal film was made an electrode in an electrochemical cell, and a potential applied, large changes in reflectance were observed. These were the beginnings of a fruitful coupling of two fields, internal reflection spectroscopy and electrochemistry.

The above observations were discussed with various electrochemists, creating considerable interest. However, not until 1966 was further work published. Mark and Pons[2] showed that, using a germanium plate as both the internal reflection element and electrode, the ATR spectrum of a saturated 8-quinolinol solution in dimethyl foramide containing 0.1 *M* $LiClO_4 \cdot 3H_2O$ as supporting electrolyte, was changed by electrolysis. Hansen, Kuwana, and Osteryoung[3,4] published a relatively comprehensive account of a method they developed for the observation *in situ* of electrochemically generated species. They used their method to study the electrochemical oxidation of o-tolidine. They found their results to be quantitatively consistent with optical, chemical, and electrochemical theory.

Work on reflection spectroscopy coupled with electrochemistry is continuing at the University of Michigan under H. B. Mark, at Case Institute of Technology under T. Kuwana, and at North American Aviation Science Center. Mark and coworkers[5] have shown that platinum films can be used as internal reflection electrodes. Kuwana and Srinivasan[6] have studied, both theoretically and experimentally, the use of tin oxide coated glass plates as internal reflection elements. In recent research[7] gold films have proved to be very successful as IRS electrodes.

THEORY

A useful internal reflection spectroscopy (IRS) - electrochemical method must meet the following criteria: (a) The reflectance must be sensitive to some electrochemical change of interest. (b) The change in reflectance must be measurable with reasonable convenience. (c) It must be possible to identify the cause of an observed reflectance change in terms of electrochemistry.

First consider the problem in a rather general form. Suppose the IRS electrode is planar and that the electrode and surrounding phases represented as a stratified medium, *i.e.*, a medium whose optical properties vary only along one coordinate normal to the interfacial planes, *i.e.*, it comprises a stack of homogeneous layers. Certain aspects of the optics of a general stratified medium are presented in an elegant way by Abeles[8]. A discussion of this case, addressed more directly to the problem at hand, is given by Hansen.[9]

From the general theory certain points are especially helpful to keep in mind. (a) Any homogeneous region can be characterized by refractive index, n, attenuation index, $\varkappa$, and thickness h. A change in reflectance of a stratified medium means that one or more of the layers has changed in n, $\varkappa$, or h, or a new layer has been formed. By proper optical design, the reflectance can be made sensitive to any given one of the optical parameters of a given film. (b) From the generalized form of Snell's law, the complex angle of refraction in the final (semi-infinite) phase of the stratified medium is independent of all phases except the incident and final phases. This is particularly important here because the angle at which there is no refracted beam in the final phase, *i.e.*, the critical angle, is independent of the presence of intermediate layers. For the same reason the depth of penetration of radiation in the final phase is independent of intermediate phases. When working at greater than critical angles, any light lost, *i.e.*, any deviation from a reflectance of unity, is due to absorption in some phase. (c) The rate of energy absorption at any point is proportional to $n^2\kappa\langle E^2\rangle/\lambda_o$ at that point. Here $\langle E^2\rangle$ is the mean square electric field and n and κ refer to the phase being considered. λ_o is wavelength *in vacuo*. This is an extremely important point; when the field distribution is known, this theorem gives physical insight into why the reflectance changes, by showing where the energy is dissipated. (d) The transverse components of $\langle E^2\rangle$ are continuous across interfacial planes, while the normal components are related by $(n_i^2+n_i^2\kappa_i^2)^2\langle E^2\rangle_i = (n_f^2+n_f^2\kappa_f^2\langle E^2\rangle_f$. For a metal with n=10 and κ=3, $\langle E\rangle$ squared normal is lowered by a factor of one million in passing from air into metal. (e) As films become very thin their presence ceases to have appreciable effect upon the surrounding mean square fields. Therefore, if the fields can be ascertained for the simpler film-absent case, they are known approximately for the film-present case. From the known fields in a thin absorbing layer (an electroadsorbed molecular layer, for example) it is easy to calculate the expected change in reflectance.

Some interesting results of a detailed analysis of stratified media with only a few phases are given below.

The optical theory of the two phase case has been extensively treated in works on attenuated total reflection. This case obtains only if the incident phase is itself the electrode. An example is a germanium ATR prism. In addition, the second phase, *i.e.*, the electrolyte, must be substantially optically homogeneous throughout the penetration depth of the radiation, *i.e.*, the inhomogeneities must have no appreciable optical effect. When these conditions are met, transmission-like spectra can readily be obtained. Since transparent electrically conducting plates or prisms are few, most actual cases are not two-phase. When they are, we can expect the usual ATR treatment of sensitivity, Beer's law behavior, etc., to apply.

The simplest type of three phase case is a perturbation on the two phase case by adding a thin layer at the interface.

This corresponds, for example, to electroadsorption of a molecular layer from the electrolyte onto the electrically conducting plate. The radiation fields that the molecular layer will see are those predicted by two-phase ATR calculations, and it can be shown[9] that a transmission-like spectrum will result.

Practical IRS-electrochemical situations will usually be more complex. The IRS-electrochemical element itself will usually comprise at least the incident phase and the thin film electrode supported by it. There may also be additional phases such as a layer to bond the electrode to the incident phase, or multilayer electrodes. Two cases have been analyzed. One case is typified by tin oxide coated glass. The other is typified by a thin metal film on a transparent incident phase.

The case of a relatively thick slightly absorbing electrode on a dielectric, such as tin oxide on glass, is discussed in detail in reference (4). The electrode thickness is of the order of a few quarter-wave lengths, and its $\kappa \ll 1$. The convenient way to make spectrophotometric measurements of reflection absorbance (or reflectance) is to take as zero, readings for some convenient state of the electrolyte, such as clear solution. A key to understanding the optical behavior of this type of IRS element, when an absorption-like spectrum of electrochemically produced species is desired, is that the small κ of the electrode, to a first approximation, does not affect the spectrum of anything in the electrolyte. The presence of the electrode layer certainly does have an effect, as shown by figure 5 of reference (4), if its refractive index is different from its substrate. However, this effect can be calculated ahead of time as indicated. An important consideration is that the sensitivity of the system to absorbing species in the electrolyte is a sensitive function of wavelength, angle of incidence, and the thickness of the electrode. This is a complicating factor in spectral interpretation, but is an advantage when high sensitivity is desired to detect absorbing species, *i.e.*, electrode thickness and angle of incidence can be adjusted to give maximum sensitivity over a limited wavelength range of interest.

We are currently investigating thin metal films as IRS-electrodes. Theoretical calculations show some interesting features, some of which were surprising. For parallel polarization at angles of incidence considerably larger than critical: (a) Very thin metal films give spectra dependent essentially only on κ of the electrolyte and not on small changes in n due to dispersion. By "spectra" we mean the change in absorbance caused by a change in the electrolyte.

(b) The intensity of the spectra taken with the metal film present may be enhanced, rather than decreased, relative to an ATR spectrum taken using the metal-free dielectric substrate. (c) Using the optical constants of gold in the visible spectral range, it is found by calculation that the sensitivity to the κ of the electrolyte is essentially constant with wavelength.

The above features are shown by numerical calculations made with the aid of a computer and equations of reference (9). They are surprising in light of the fact that a large fraction of the incident light energy may be absorbed by the metal film itself, and the optical constants of gold, for example, change rapidly throughout the visible region.

EXPERIMENTAL

The electrochemical problem at hand dictates the basic experimental design. Detailed developments must involve finding materials which are optically and electrochemically suitable for IRS-electrochemical electrodes. The systems used to date are the multiple ATR germanium plate used by Mark and Pons[(2)] and systems similar to the one shown in Figure 1 used by Hansen *et. al*. Various types of IRS-electrode systems can be used with this same basic cell design, with various degrees of complexity in the resultant spectra.

A cell of the type shown in Figure 1 was used by Hansen, Kuwana, and Osteryoung in their study of the oxidation of o-tolidine in aqueous solution. In this case the IRS element was tin oxide coated glass. The tin oxide which has an index larger than that of glass was thick enough to give large interference effects. This gave rise to a complicated, but understandable, reflection spectrum. The sensitivity of the system was also a function of wavelength. At a given wavelength, however, the thickness of the tin oxide layer, or the angle of incidence could be chosen to give maximum sensitivity. The electrochemical manipulations used included potential sweep, potential step, chronopotentiometry, and potentiometry. The results agreed quantitatively with theory. In particular, they showed (a) that the species "observed" were indeed those within a "penetration depth" of the electrode surface, *i.e.*, within about 500 A distance; (b) the diffusion equations of electrochemistry predicted accurately the observed changes in surface concentration with time and potential; (c) that the observed reflectance behavior was entirely consistent with our optical analysis. For example, at fixed wavelength

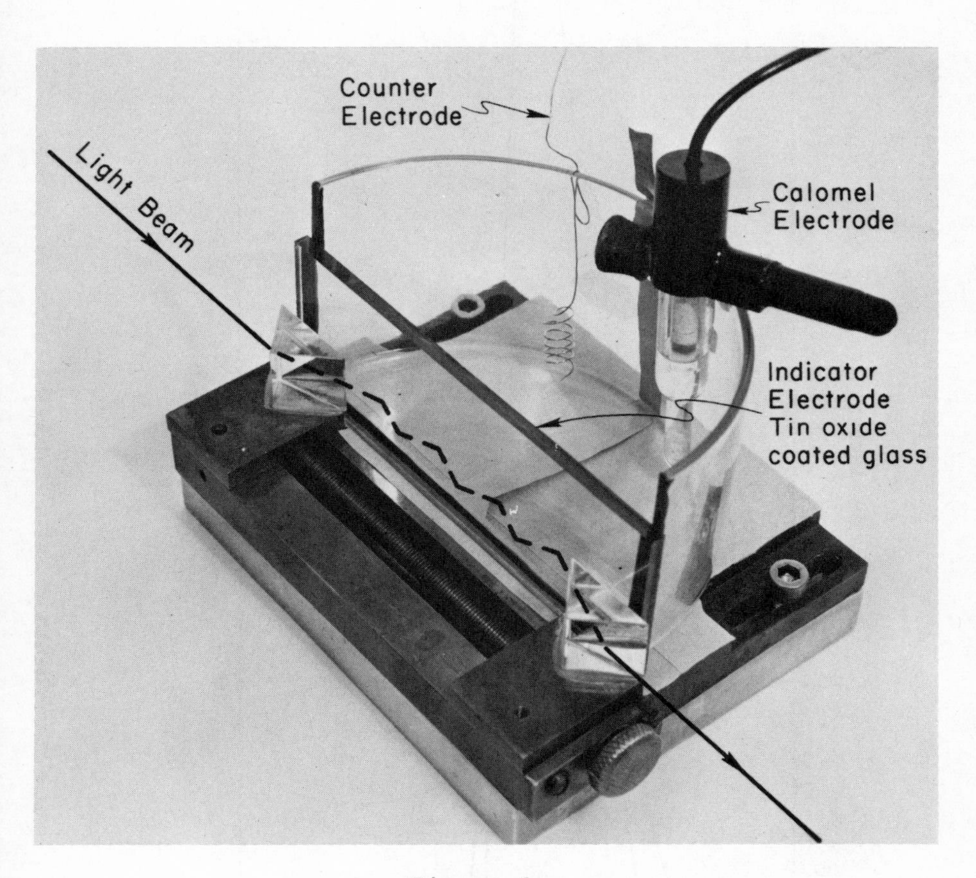

Figure 1.

IRS-electrochemical cell for observing electrochemical phenomena spectrophotometrically, *in situ*.

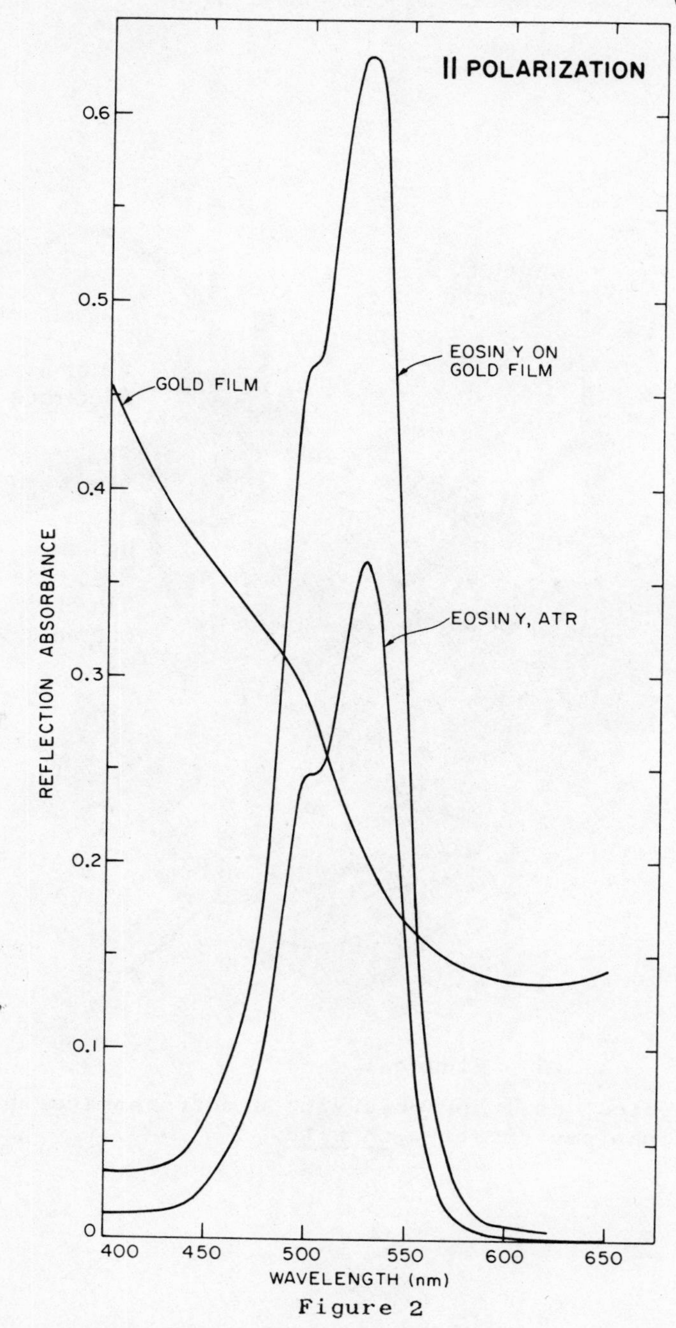

Figure 2

Internal reflection absorbance spectra of a thin gold IRS electrode on glass, eosin Y solution on glass, and eosin Y solution on gold on glass, for parallel polarization.

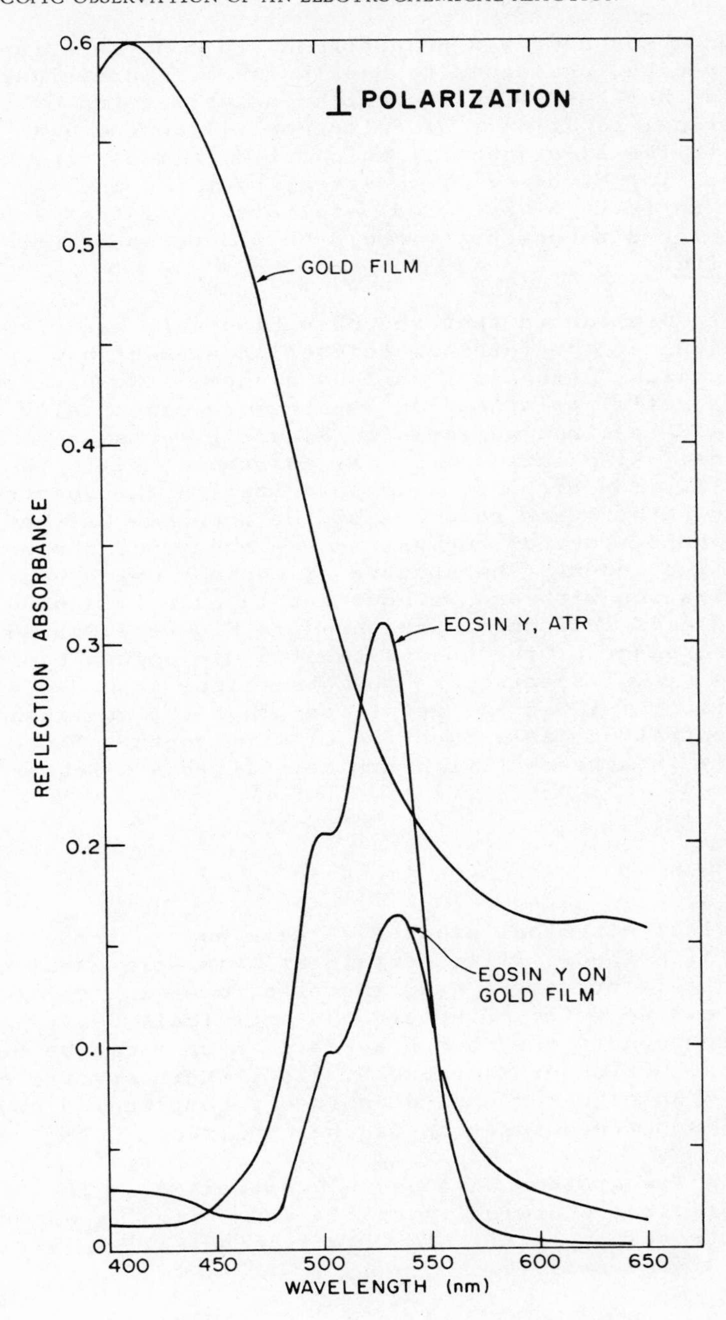

Figure 3

IRS spectra as in Fig. 2, except for perpendicular polarization.

reflection absorbance was proportional to concentration of colored species, analogous to Beer's law. A more significant example is that the effective thickness calculated by the purely optical equation 7 of reference (4) agrees quantitatively with the same quantity calculated from electrochemical data (potential step and chronopotentiometric) and the known molar absorptivity of oxidized o-tolidine; (d) that a metastable, colored intermediate could be studied spectroscopically *in situ*.

A cell similar to that shown in Figure 1, but with thin gold on glass as the internal reflection element has been used by Prostak[7] to study various systems. He has found that for parallel polarization spectra are essentially enhanced ATR spectra, as shown in Figure 2 for aqueous eosin Y solutions. Also included is the absorbance of the gold film itself. For perpendicular polarization the spectra are diminished rather than enhanced by the presence of the gold film and are distorted somewhat by the anomalous dispersion of refractive index. The spectra of eosin Y using perpendicular polarization with and without a thin gold film on glass, at 72° angle of incidence, are shown in Figure 3. Also given is the absorbance of the gold film at a dye concentration of zero. The eosin Y spectra are the absorbance with dye at a concentration of 100 grams per liter minus the absorbance at zero concentration, as a function of wavelength. These results are in agreement with our calculations discussed above.

CONCLUSIONS

IRS-electrochemical studies to date have introduced a new and powerful method. It is certain at this point that changes in very thin layers at or near the electrode-electrolyte interface can be detected spectrophotometrically. In fact, under proper conditions, the changes in n or $\varkappa$ can be determined as a function of wavelength, *i.e.*, their spectra can be obtained. The required apparatus is not complicated and commercial spectrophotometers can be utilized.

Only a few systems have been investigated so far. When a number of electrochemically stable and optically understood electrodes are available, electrochemical research can proceed more readily.

REFERENCES

1. W. N. Hansen, L. Lynds, and R. A. Osteryoung, "Infrared Reflectance Study of Gas-Solid Interaction," Final Summary Report, October 1962, AD-298 877 L; Proc. of ARPA Fuel Cell Conf., Whiting, Ind., February 1962.
2. H. B. Mark and B. S. Pons, Anal. Chem., 38:119 (1966).
3. W. N. Hansen, R. A. Osteryoung, and T. Kuwana, J. Am. Chem. Soc., 88:1062 (1966).
4. W. N. Hansen, T. Kuwana, and R. A. Osteryoung, Anal. Chem., 38:1810 (1966).
5. B. S. Pons, J. S. Mattson, L. O. Winstrom, and H. B. Mark, Anal. Chem., 39:685 (1967).
6. V. S. Srinivasan and T. Kuwana, "Internal Reflectance Spectroscopy at Optically Transparent Electrodes," J. Phys. Chem., in press.
7. Arnold Prostak, private communication.
8. F. Abeles, Annales de Physique, 5:596 (1950).
9. W. N. Hansen, "Electric Fields Produced by the Propagation of Plane Coherent Electromagnetic Radiation in a Stratified Medium," J. Opt. Soc. Am., in press.
10. N. J. Harrick, "Internal Reflection Spectroscopy," Interscience Publishers, New York, 1967.

THREE UNUSUAL APPLICATIONS OF INTERNAL REFLECTION SPECTROSCOPY: IDENTIFICATION OF GC FRACTIONS, PYROLYSIS AND SKIN ANALYSIS

PAUL A. WILKS, JR.

WILKS SCIENTIFIC CORPORATION

SOUTH NORWALK, CONNECTICUT

Previous papers have discussed the theory and general applications of internal reflection spectroscopy. This paper will deal with three specific applications for which special equipment has been developed.

The first case is the trapping and identifying of gas chromatic fractions. The problem here is the efficient extraction of a small amount of material from a relatively large amount of carrier gas into a sensitive infrared absorption cell.

The multiple internal reflection plate can provide a highly sensitive method for obtaining spectra from a thin layer of molecules deposited or condensed along its surface. In multiple internal reflection, the beam of radiation reflects internally a number of times as it travels the length of the plate, penetrating into the sample with each reflection. Since the penetration depth is of the order of a wavelength of light, the beam may effectively transmit through the sample during each reflection. The additive effect of the large number of internal reflections provides a well defined spectrum from a very small volume of material.

A volatile material deposited on the surface of an internal reflector plate would quickly evaporate. However, if a second plate is placed over the first, the material would be held as a capillary film. Figure 1 shows such an arrangement schematically. Note that where each plate is half the normal thickness, twice the number of reflections take place. Since half the energy goes down each plate, the net effect is to double the sensitivity when sample is placed all around the two plates or provide equal sensitivity when sample is placed between them only.

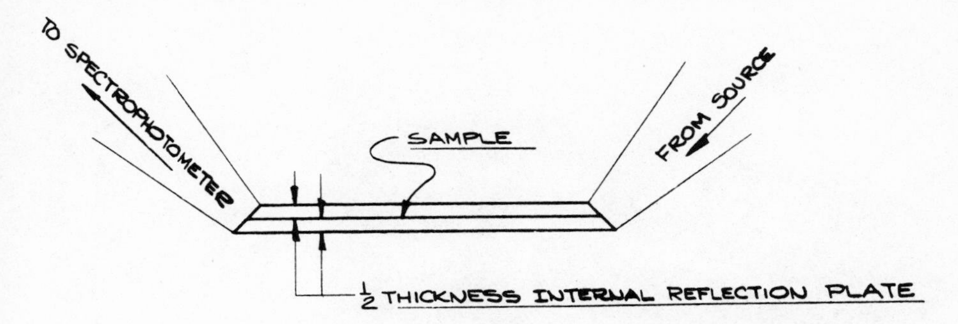

CAPILLARY INTERNAL REFLECTION CELL *

* PATENT APPLIED FOR

Figure 1

To utilize the capillary internal reflection cell in collecting GC fractions, the cell is mounted so one face is in contact with a thermo-electric cooling element (figure 2) and the cell opened at one end slightly so that the hot carrier gas pushes across it (figure 3).

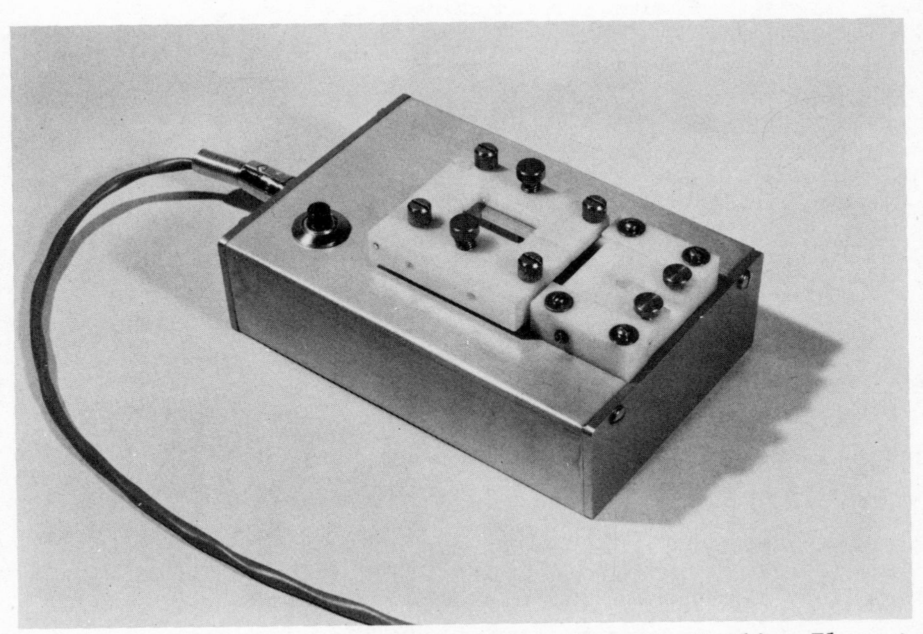

Figure 2 Cell In Contact With Thermo-Electric Cooling Element

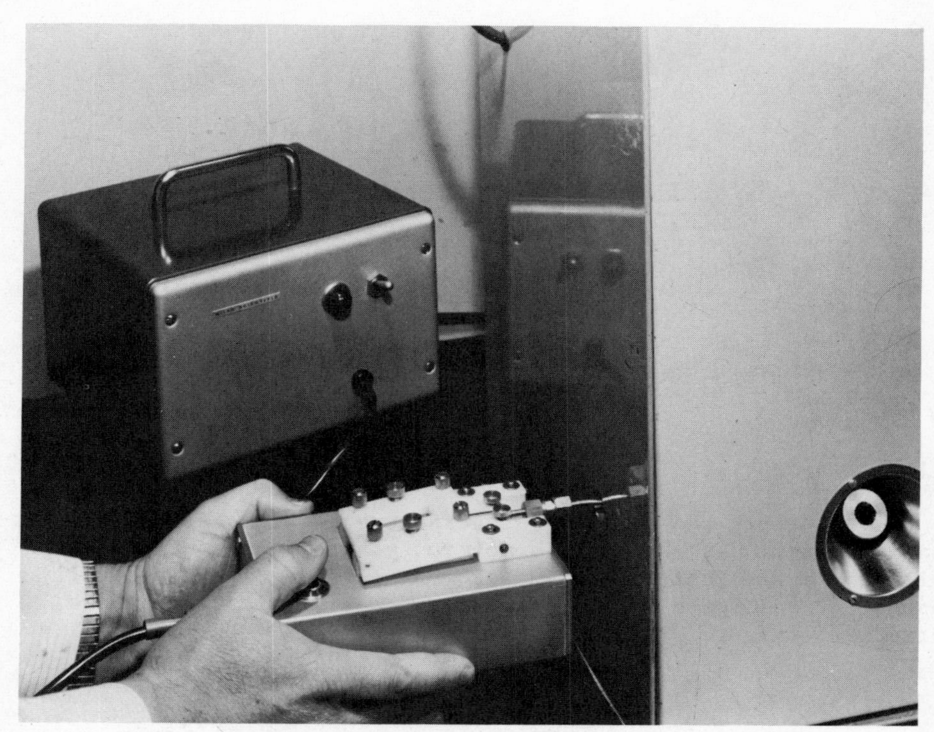

Figure 3 Collecting A Fraction

Figure 4 Cell Mounted In Spectrophotometer

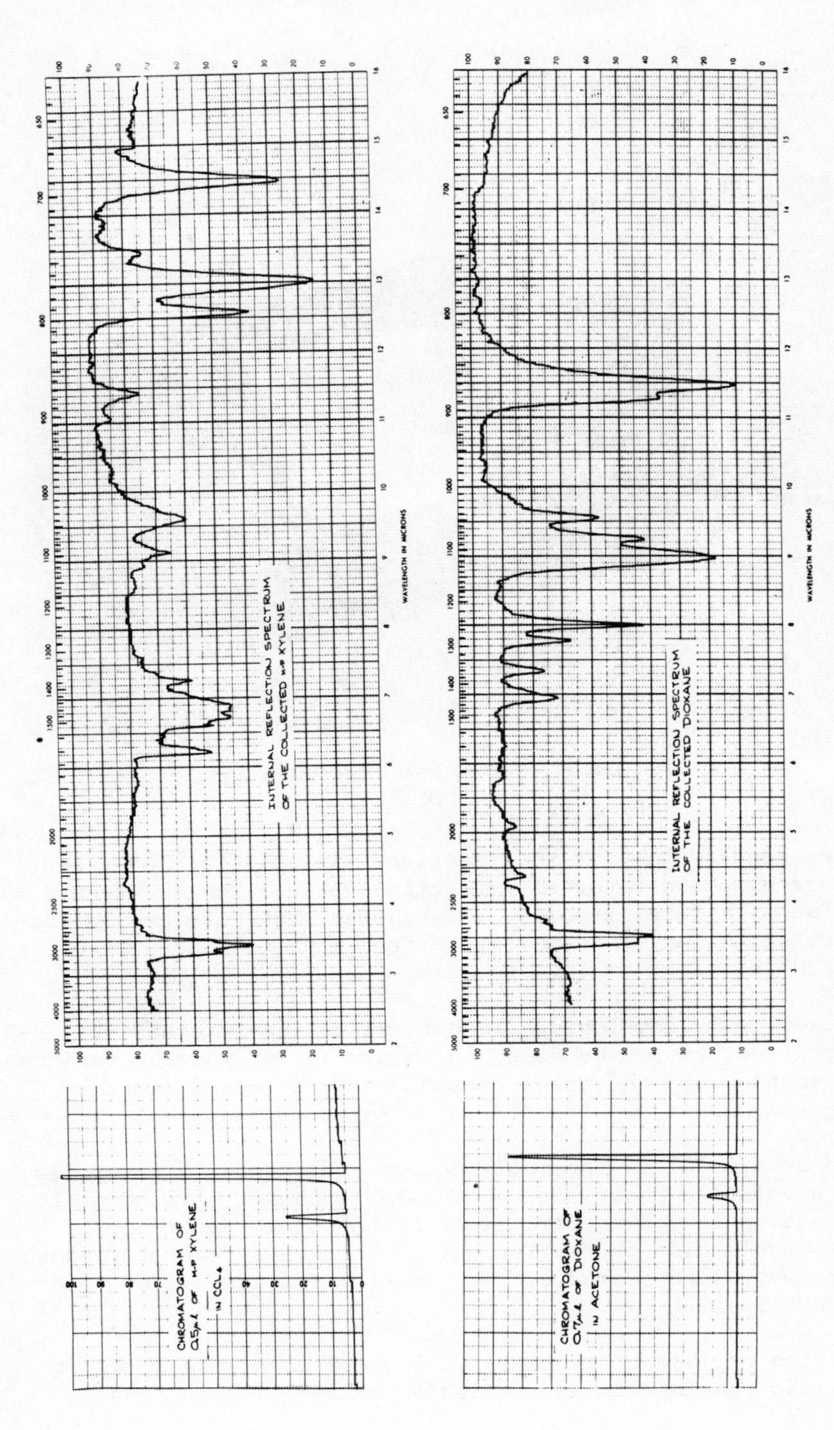

Figure 5 Spectra Of The Collected Fractions

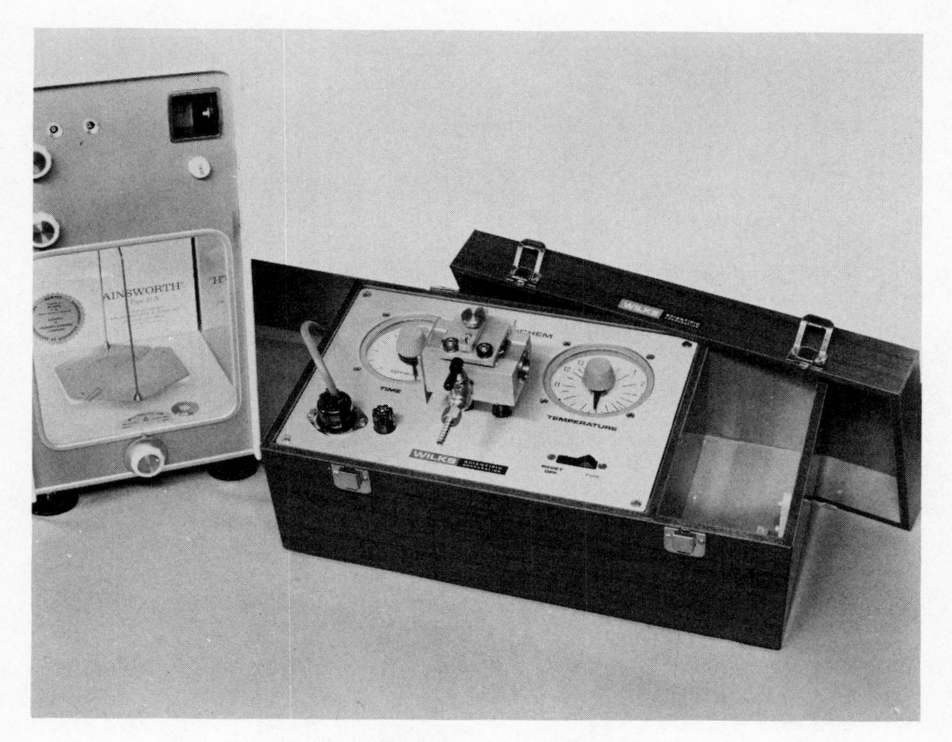

Figure 6 Pyro-Chem Accessory

After condensation of the fraction, the cell is closed and placed in a standard internal reflection optical system (figure 4). Figure 5 shows typical results. In general, the internal reflection fraction collector is useful on fractions boiling between 75°C and 200°C and .1 to .5ul in size.

Another newly developing field in which the internal reflectance technique has proved valuable is pyrolysis, whereby various types of samples are thermally degraded or decomposed and the resulting components are identified by infrared analysis.

When a material is thermally decomposed, there is generally produced a volatile phase and a condensable phase. For complete analysis, the infrared spectra of both phases are desirable. Pyrolysis should also be carried out in an oxygen free atmosphere and under controlled reproducible conditions of temperature and heating duration.

A device capable of meeting these requirements is shown in figure 6. It consists of an evacuable chamber with rock salt windows on each end and an internal reflector plate on the top side. A nickel-chrome firing filament on which the sample is placed is positioned in the bottom of the chamber opposite the internal reflector plate. The chamber plugs into a power supply which includes controls for setting temperature and an adjustable timer.

After firing, the chamber is moved to the infrared spectrophotometer where it is inserted in the normal cell holder for analysis of the vapor phase (figure 7) and in a standard internal reflection optical system for analysis of the condensed phase (figure 8).

Figure 7 Pyrolysis Chamber Mounted To Record The Spectrum Of The Vapor Phase

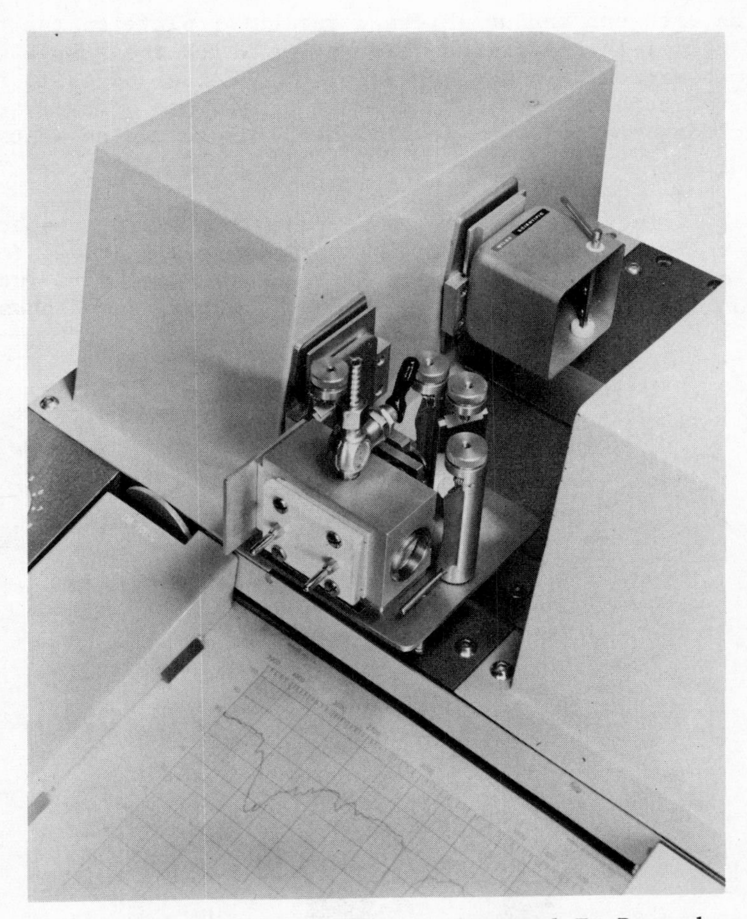

Figure 8 Pyrolysis Chamber Mounted To Record
The Spectrum Of The Condensed Phase

The applications of the pyrolysis-infrared technique are the following:

1. Identification of additives and plasticizers.

Slow gentle beating of a sample will often cause additives and plasticizers to vaporize so that they can be identified. After infrared analysis the chamber can be re-evacuated and higher temperatures used to thermally decompose the sample.

2. Elimination of fillers.

Plastics are often filled with such materials as carbon

black or clay which make spectral identification difficult or impossible. Pyrolysis will usually effect a separation permitting the identification of the polymer.

3. Thermal Decomposition.

Polymeric materials will degrade into simpler molecules in three ways under heat:

1. Chain scission - where the polymer chains simply break into smaller segments.

2. Unzipping - where major groups split off from the polymer building blocks.

3. Thermal Decomposition - where the molecule breaks down into basic groups.

Figure 9 is Nylon 66. The sample was fired at 800°C for one minute. The spectrum at the top of the figure if of the vapor phase materials evolved. The sample has undergone some decomposition evidenced by the ammonia present in this vapor phase spectrum. The center spectrum is of the materials condensed on the MIR plate. It is clearly a polyamide although the spectrum is changed from the MIR spectrum of the Nylon 66 staple (bottom spectrum).

Figure 10 illustrates the unzipping and chain scission reactions possible with pyrolysis. A polystyrene fiber sample was fired at 800°C for a relatively long term (four to five minutes). Styrene monomer can be seen in the vapor phase spectrum (top) while the center spectrum is almost identical to the MIR spectrum of polystyrene. The band at 780 cm^{-1} is probably due to a change in substitution on the aromatic ring.

Figure 11 is typical of polymers containing nitrile groups. Orlon and Creslan illustrated here are both polyacrylonitrile fibers. The vapor phase spectra contain essentially one band at about 715 cm^{-1} due to hydrogen cyanide. The condensed phases (center spectra) are similar to the MIR spectra of the fiber with possibly some absorption from H_2O.

Figure 12 includes spectra of the pyrolysis products of an acrylonitrilebutadiene elastomer and again shows the characteristics of nitrile containing materials. The vapor phase contains hydrogen cyanide (715 cm^{-1}), 1,3-butadiene plus a trace of ethylene and methane. The condensed phase spectrum (bottom) is similar to the MIR spectrum of the unfilled material.

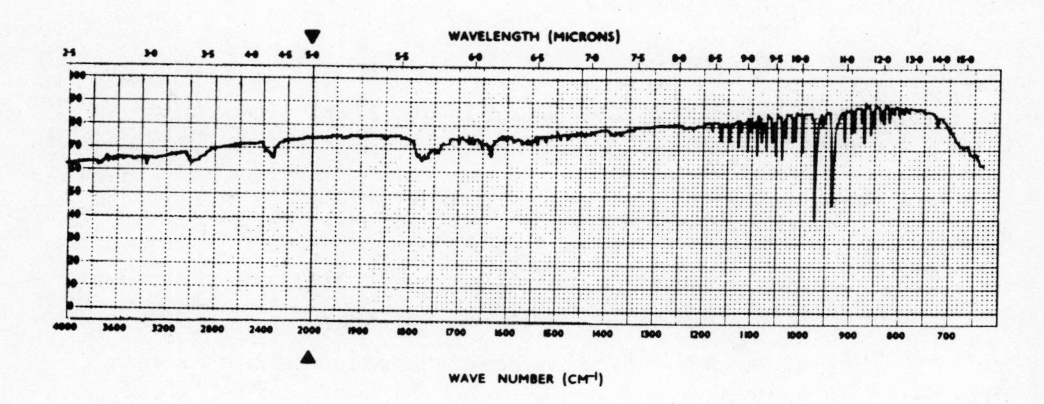

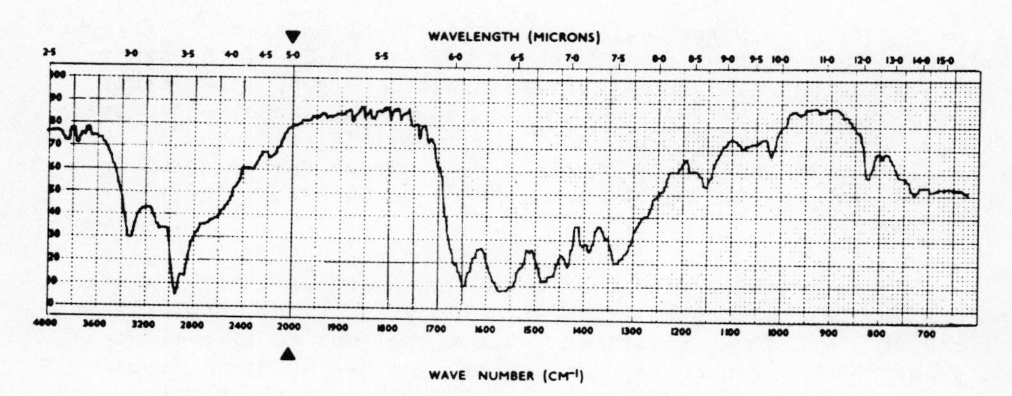

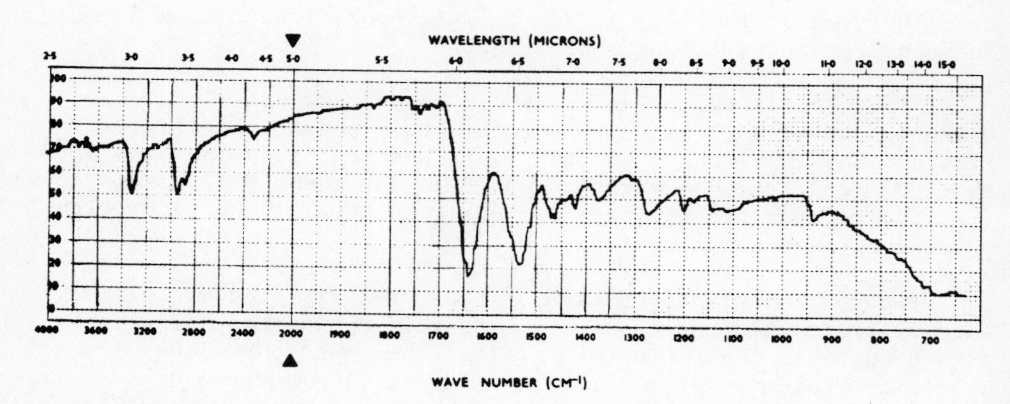

Figure 9 Spectra Of The Pyrolysis Products Of Nylon 66

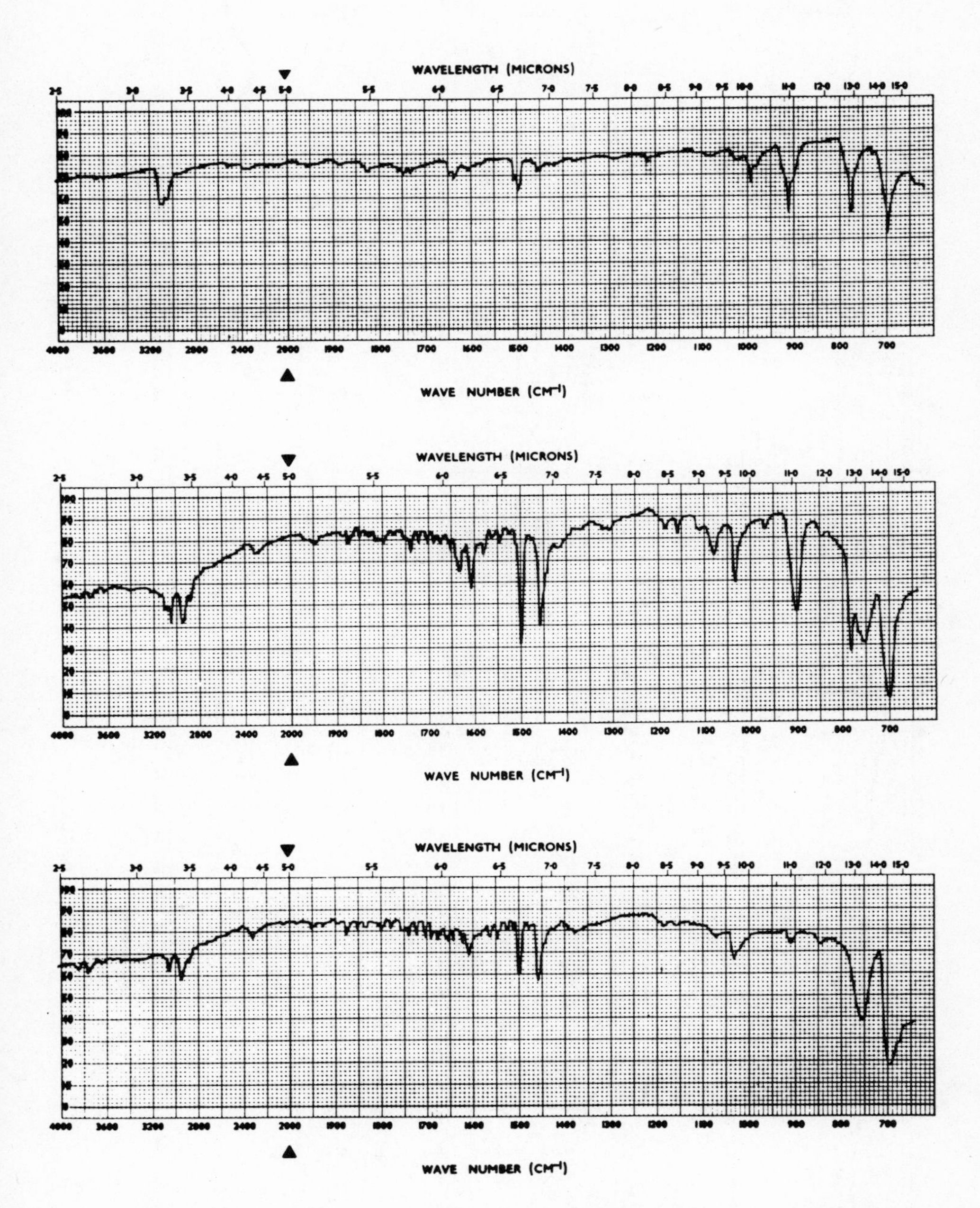

Figure 10 Spectra Of The Pyrolysis Products Of Polystyrene

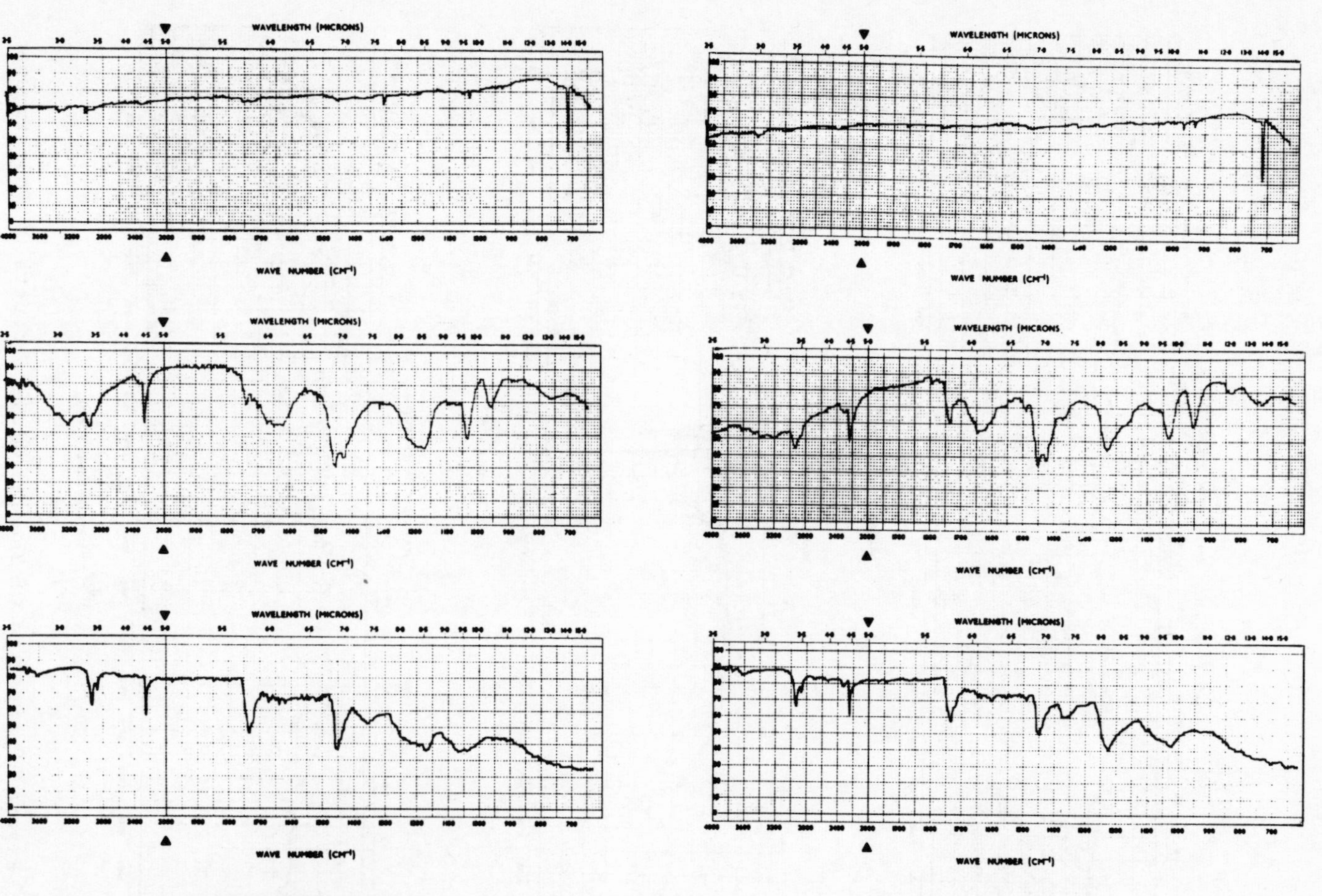

Figure 11 Spectra Of The Pyrolysis Products Of Orlon And Creslan

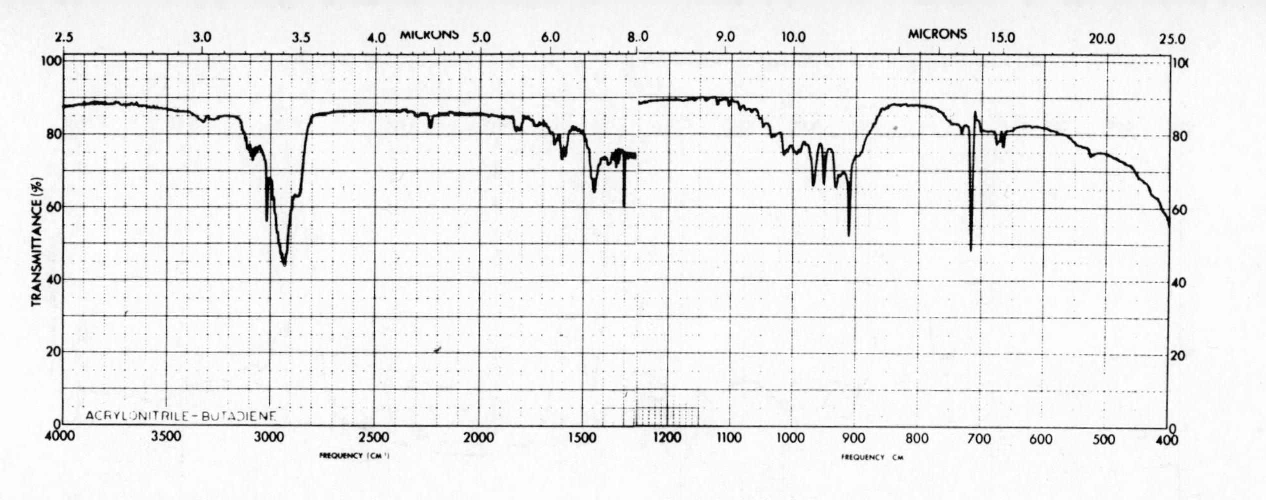

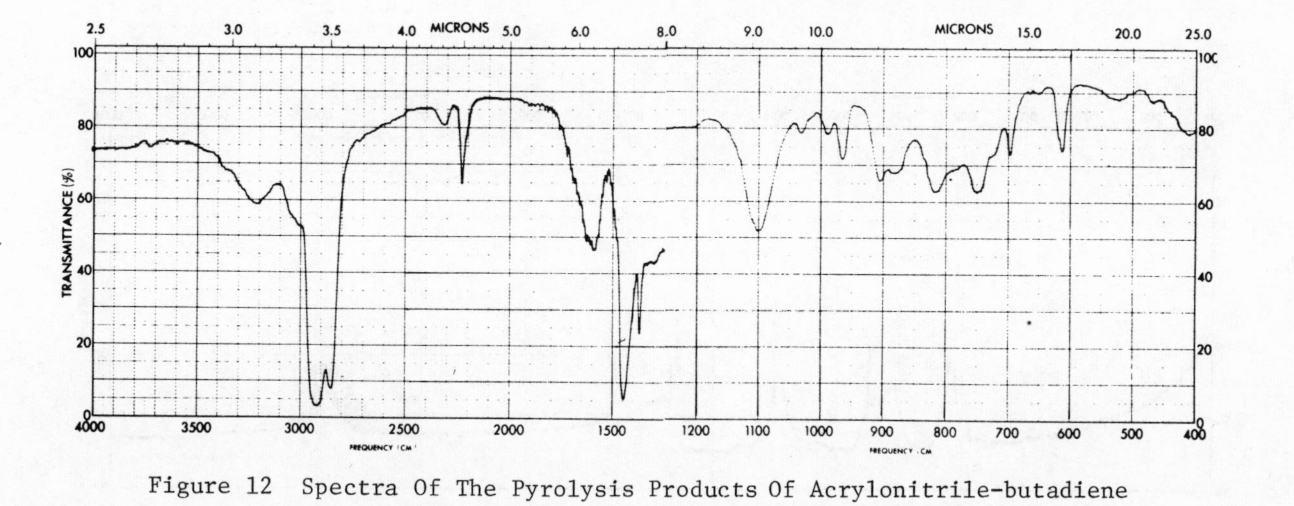

Figure 12 Spectra Of The Pyrolysis Products Of Acrylonitrile-butadiene

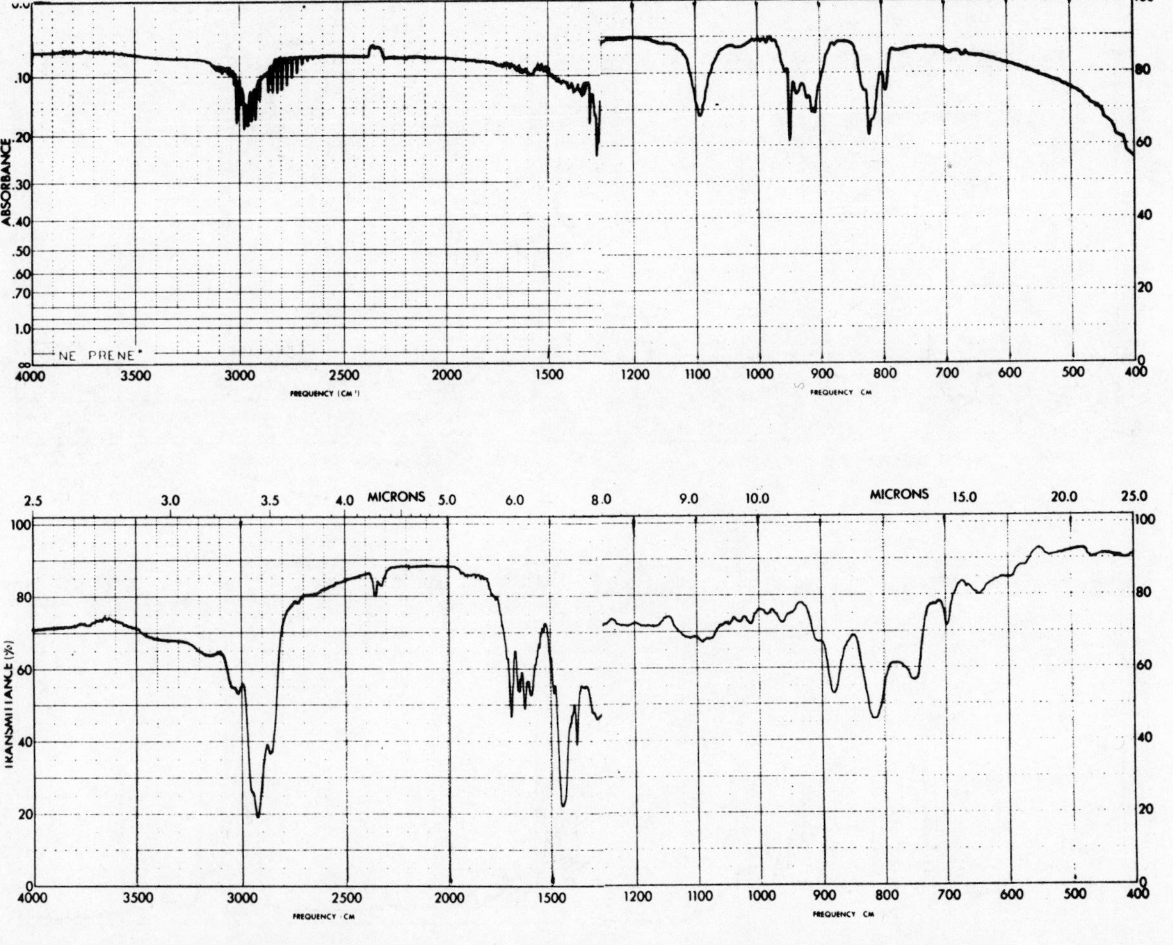

Figure 13 Spectra Of The Pyrolysis Products Of Neoprene

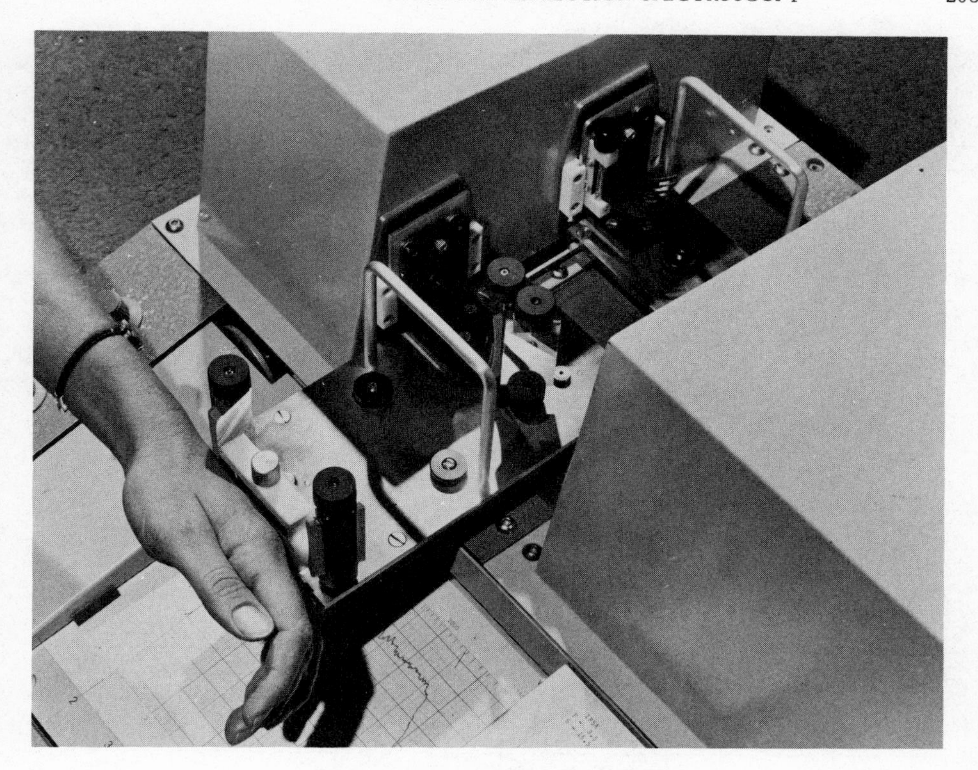

Figure 14 Skin Analyzer

Figure 13 is of Neoprene. The vapor phase contains HCl plus a mixture of chlorinated materials, possibly perchloroethylene-vinylidene chloride and related materials. The condensed phase is different from the MIR spectrum of the original polymer.

And finally, it is possible through the proper arrangement of optical components to bring a sensitive surface of an internal reflector plate beyond the sample compartment of a spectrophotometer so that absorption spectra can be obtained on relatively large samples. An example is the in situ analysis of skin.

A typical arrangement is shown in figure 14. Note that nearly any part of the anatomy can be brought in contact with the reflector plate. Useful information can be obtained on the retention times of certain material on the skin. Studies are underway to determine whether or not infrared spectroscopy can be used as a diagnostic tool in dermatology. The spectra in figures 15 and 16 show "Clean" skin and skin to which a lotion has been applied.

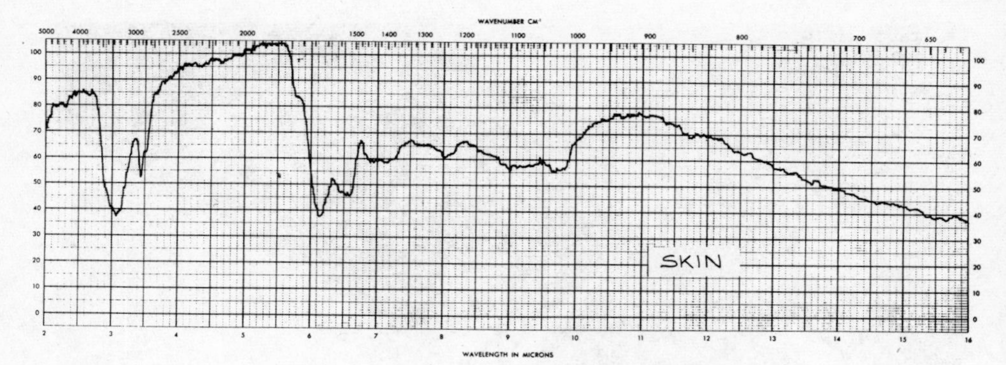

Figure 15 "Clean" Skin

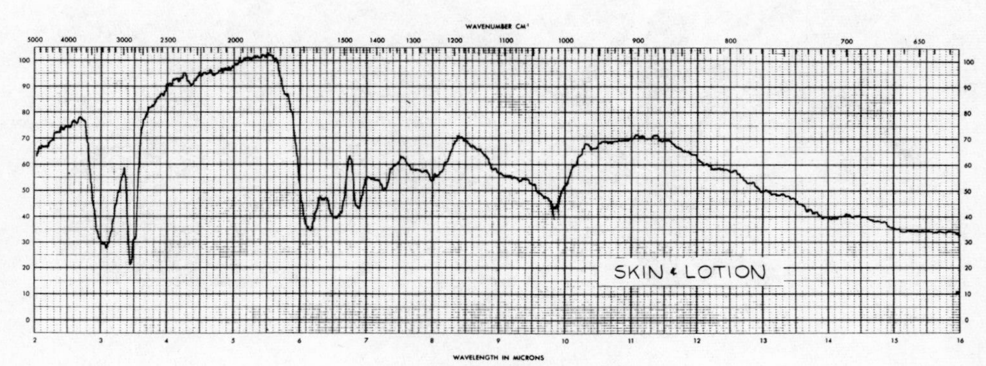

Figure 16 Skin + Lotion

Another interesting application of this optical arrangement is in the spectral study of large objects that cannot be fitted into conventional infrared cells. Spectra have been obtained on a number of paintings leading to the interesting possibility that infrared analysis can help authenticate paintings.

Thus, the internal reflection technique gives a new dimension to infrared spectroscopy, adding considerably to the versatility of an already highly versatile analytical method.

ENHANCED SENSITIVITY FOR INTERNAL REFLECTION SPECTROSCOPY

N. J. Harrick

PHILIPS LABORATORIES

Briarcliff Manor, New York

ABSTRACT

A standing wave with an evanescent electromagnetic field in the rarer medium is established normal to the reflecting surface for total internal reflection. Optical spectra of materials can be obtained by measuring the interaction of this evanescent field with an absorbing rarer medium. This spectroscopy technique, called Internal Reflection Spectroscopy, can be employed in many instances where conventional either fail or cannot easily be employed. Examples of these are the recording of the spectra of monomolecular films and powdered samples. By placing appropriate resonant thin films (optical cavities) on the reflecting surface, strongly amplified fields, analogous to that obtained in microwave cavities are established, and enhanced absorptions are obtained. In addition to a brief review of Internal Reflection Spectroscopy, the principles and applications of these optical cavities will be discussed as well as other methods for enhancing the sensitivity.

INTRODUCTION

It is well-known that when light is internally reflected from the surface of an optically transparent medium at angles exceeding the critical angle, total reflection occurs. The superposition of the incoming waves with the reflected waves form a standing wave normal to the surface in the vicinity of the reflecting surface as shown in Fig. 1. There is a sinusoidal variation of the electric-field amplitude with distance from the reflecting surface in the optically denser medium and an evanescent field whose amplitude decays exponentially with distance from the surface in the rarer medium. Optical spectra of materials can be recorded by measuring the interaction of this evanescent field

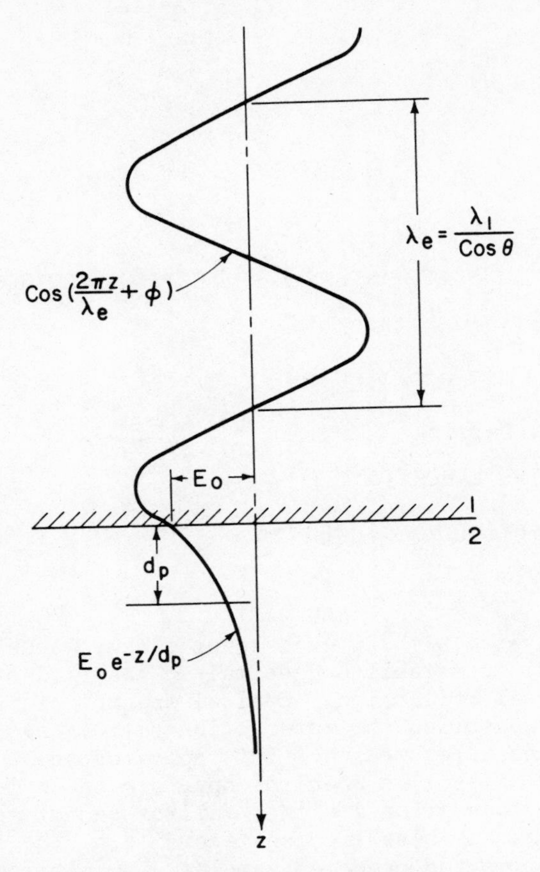

Fig. 1. Standing-wave amplitudes established near a totally reflecting interface for light incident and reflected at angle equal to θ. There is a sinusoidal dependence of the electric-field amplitude on the distance from the surface in the denser medium 1 and exponentially decreasing amplitude in the rarer medium 2. The E-field may have a large value at the surface.

with an absorbing rarer medium. This method of recording spectra is called Internal Reflection Spectroscopy (IRS)*.

There are a number of advantages that this spectroscopy technique has over conventional techniques. One advantage is the ease of sample preparation - it is only necessary to bring the sample material, be it a liquid, powder or irregular solid in the region of the evanescent field to record its spectrum. Another

*The principles of Internal Reflection Spectroscopy, complete bibliography and details of many of the points mentioned here are treated fully in "Internal Reflection Spectroscopy" by N. J. Harrick (Interscience Publishers, Div. J. Wiley &Sons, N.Y. (1967).

advantage is that the bothersome interference fringes associated with propagating waves and present in transmission measurements are absent. These fringes tend to obscure weak absorption bands and contribute to uncertainties in making absolute measurements. Interference fringes are absent in internal reflection measurements because the evanescent field is a non-propagating wave.

The degree of interaction of the evanescent field with an absorbing rarer medium can be determined precisely from Maxwell's or Fresnel's equations. These exact mathematical calculations, however, give little insight into the nature of the interaction of the evanescent field with the absorbing rarer medium. It is therefore informative to consider the low absorption approximation of these equations where each of the factors which determines the strength of interaction is clearly evident. The reflection coefficient of the interface can be written as

$$R = 1 - \alpha d_e \quad . \qquad (1)$$

Here α is the absorption coefficient. d_e is the effective thickness and is equal to the thickness required to yield the same absorption loss in a transmission measurement. This effective thickness is a measure of the strength of interaction of the evanescent wave with the absorbing medium. Simple expressions for two limiting cases, viz., bulk materials and very thin films, can be written and these will be discussed here.

1. Bulk Materials

For bulk material the effective thickness is given by

$$\left.d_e\right)_B = \frac{n_{21} E_o^2 d_p}{2\cos\theta} \quad . \qquad (2)$$

Here n_{21} is the refractive index ratio of the interface, E_o is the electric-field amplitude at the interface for an incident wave having unit electric-field amplitude and d_p is the depth of penetration of the evanescent wave in the rarer medium. This equation shows that there are four factors which determine the strength of coupling of the evanescent wave to the absorbing rarer medium. The net effect of all of these factors is to give a decrease of the strength of coupling with increasing angle of incidence (θ). The following four factors are involved:

a. Depth of penetration. The best known of these factors is the depth of penetration of the evanescent wave, d_p, which decreases with increasing θ. The depth of penetration is the same for both polarizations.

b. Electric field strength. Another factor which controls the strength of coupling is the electric field intensity of the standing wave at the reflecting interface, E_o^2, which also decreases with increasing θ. Here E_o is the electric-field amplitude in the rarer medium at the interface for unit incoming amplitude in the denser medium and is greater for $||$-polarization than it is for $\perp$-polarization.

c. Sampling area. The third factor is the sampling area. It is proportional to $1/\cos\theta$ and increases with increasing θ. This factor is obvious in transmission measurements in which the sample thickness increases as $1/\cos\theta$ for oblique incidence.

d. Index matching. The last factor is the matching of the refractive index of the denser medium to that of the rarer medium, given by n_{21}. Index matching controls the strength of coupling. This term is independent of θ. It predicts an increase of coupling with better matching, i.e., as $n_{21} \rightarrow 1$.

Because the electric-field amplitudes in the rarer medium are different for equal incident amplitudes of perpendicular and parallel polarization, the effective thicknesses are different for the two polarizations - the one for parallel polarization is always greater for bulk materials.

In Fig. 2 the effective thickness and the depth of penetration have been plotted versus angle of incidence for the interface, $n_{21} = 0.423$, which corresponds to $\theta_c = 25°$. It should be noted that for $\theta = 45°$, $d_{e\perp} = (1/2)d_{e||}$ (generally true at 45°) and the average effective thickness, i.e., $(d_{e\perp} + d_{e||})/2$, is about equal to the depth of penetration. For smaller angles of incidence, the effective thicknesses are larger than the penetration depth and may be equal to many wavelengths for those angles wherein the penetration depth is the major controlling factor. For larger angles, however, the effective thicknesses are smaller than the penetration depth and, because of the rapid decrease of the E-fields with θ, approach zero as θ approaches grazing incidence.

2. Thin Films

For films much thinner than the depth of penetration of the evanescent wave, the effective thickness is given by

$$(d_e)_F = \frac{n_{21} E_o^2 d}{\cos\theta} \quad . \qquad (3)$$

The same factors control the effective thickness in this case as for bulk materials except for the penetration depth which is now replaced by the actual film thickness, d. Another difference compared to bulk materials is that for thin films the effective thickness for parallel polarization may be greater or less than that for perpendicular polarization, depending on the refractive index ratio of the interface, n_{21}. The example of d_e versus θ shown in Fig. 3 indicates that d_e may be greater or less than the actual film thickness, depending on θ and polarization.

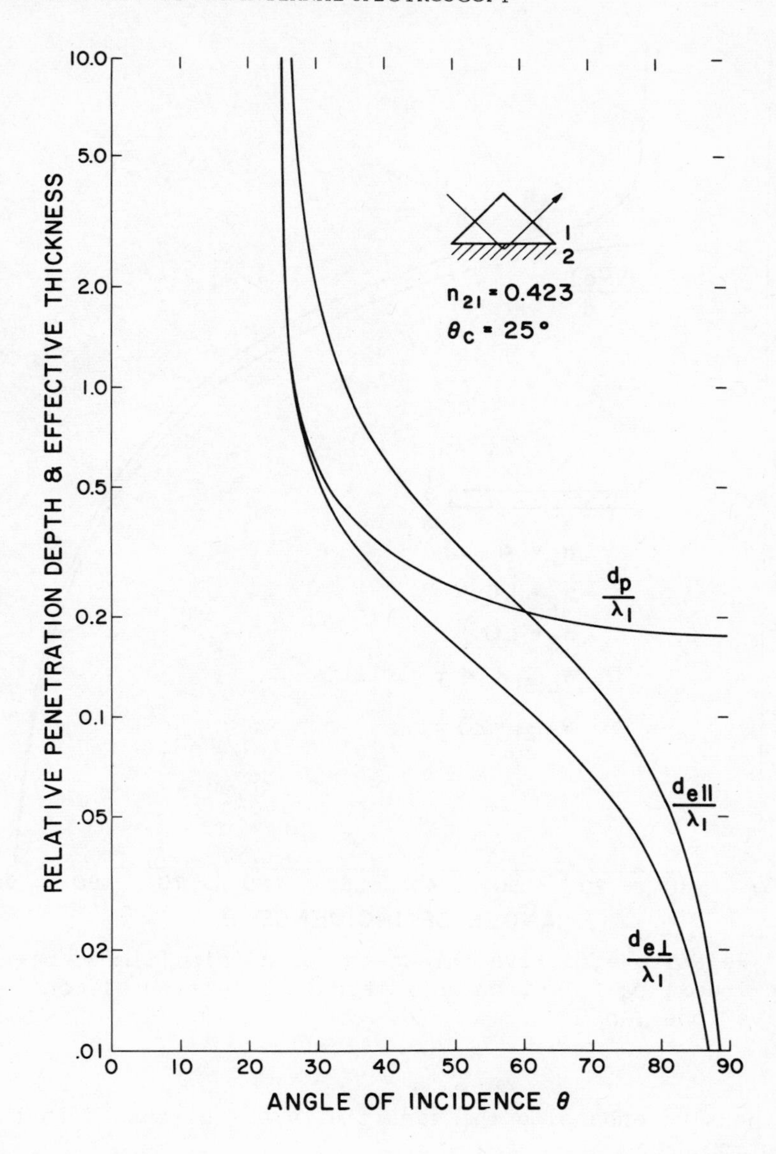

Fig. 2. Relative penetration depth of the evanescent wave and effective thickness versus the angle of incidence for an interface whose refractive index ratio is $n_{21} = 0.423$. Near the critical angle the effective thicknesses for both polarizations are greater than the penetration depth and at large angles they are less.

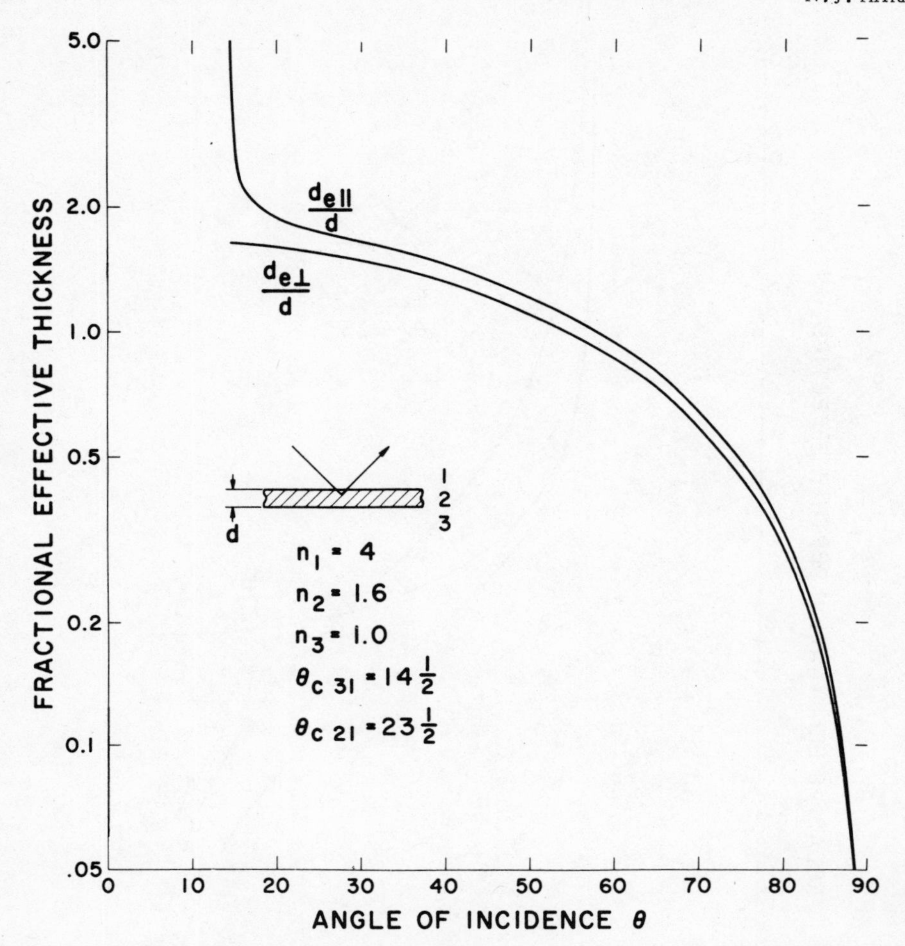

Fig. 3. Relative effective thickness for a thin film of refractive index, $n_2 = 1.6$, on an internal reflection element of refractive index, $n_1 = 4.0$.

Methods of enhancing the sensitivity are discussed in the next section.

ENHANCED SENSITIVITY

In general, sensitivity is enhanced by increasing the effective thickness by means of optimizing the parameters which control the effective thickness, by employing multiple reflections, by appropriate sample handling and by employing multiple reflections at one point for a limited quantity of sample.

1. Optimization of the Effective Thickness

The effective thickness (for bulk materials and thin films) can be optimized by matching the refractive index of the internal reflection element (IRE) to that of the sample material, i.e., $n_{21} \rightarrow 1$, and by reducing the angle of incidence to increase E_o and d_p. Spectra of fairly strongly absorbing samples can then be recorded employing a single reflection.

2. Multiple Reflections

If the sample is weakly absorbing or is thinner than a penetration depth or adequate contact between sample material and internal reflection element cannot be made, multiple reflections must then be employed to increase the effective thickness. The reflected power for N reflections is then given by

$$R^N = (1 - \alpha d_e)^N \simeq 1 - \alpha N d_e \quad . \tag{4}$$

The effective thickness is thus increased by the factor N.

If the amount of sample material is limited it must be placed on the internal reflection element so that it is efficiently sampled. For example, small liquid droplets can be distributed over the surface of the internal reflection element by capillary action via the use of another plate parallel to the surface of the internal reflection element. Spreading of liquid droplets can also be achieved by the use of a wire mesh near the surface or by finely divided powder placed on the surface of the internal reflection element. The latter methods have the advantage that they leave the surface of the IRE readily accessible.

3. Multiple Reflections at One Point

If very many reflections are required, it is advantageous to reflect the light beam more than once from a given point on the surface of the IRE since there are practical limits on the length and thickness of internal reflection plates that can be employed. Multiple reflection must be employed particularly in cases where only minute quantities of sample material are available. A number of methods have been employed to obtain multiple reflections at one point.

a. Multiple sampling plates. A few (e.g., two or four) reflections at one point can be obtained in simple internal reflection plates. This is accomplished by reflecting the beam back upon itself after it has propagated down the length of the plate. Separation of entrance and exit beams is achieved by cocking the end-reflecting surface so that the return beam then propagates at a slightly different angle of incidence than the incoming beam.

As an example of the enhanced sensitivity that can be achieved employing multiple reflections, capillary spreading and multiple sampling, a 0.01 microliter droplet of water placed on a double-sampling plate of 1.5 mm thickness resulted in absorption of about 0.5% for the O-H band at a wavelength of 2.9 microns, while the same quantity placed on a double-sampling plate of 0.4 mm thickness using powder for spreading the liquid and a light beam restricted to a width comparable to the spreading of the liquid resulted in an absorption of the infrared beam in excess of 75%! In other words, 10 micrograms of water yielded an absorption of 75%. Assuming a logarithmic law, this represents a gain of about 280 over the 1.5 mm plate. This is clearly another indication of the potential detection of submicrogram quantities via IRS.

b. Rosette. Multiple reflections from a single point can be obtained in more complicated internal reflection elements as, for example, in the rosette. The light beam is introduced into this element and is reflected many times from a single sampling point as it precesses within the structure. Structures of this type have the disadvantage of being complicated to fabricate and difficult to align.

c. Optical cavity. Resonant structures at optical frequencies, called optical cavities, show considerable promise for enhancing sensitivity for internal reflection spectroscopy. The optical cavity is a thin "Fabry-Perot" interference film employing total internal reflection. In this film many internal reflections occur within the beam width. Thus, when resonance conditions are satisfied, the multiply reflected components become superimposed to form very intense electric fields and hence strong interaction with the sample material can be obtained.

In the simplest form, the resonance condition for a half-wave film is given by

$$nt \cos\theta = \lambda/2 \quad . \tag{5}$$

Resonance is thus dependent on the index of refraction, n, and thickness, t, of the film as well as the angle of incidence, θ. The optical cavity has the disadvantage that it is substantially a fixed-frequency device although it can be tuned to some degree by changing the angle of incidence. The optical cavity is useful for detecting specific components. Figs. 4 and 5 show the enhancement that was achieved with a thin resonant film of silicon. Fig. 4 shows an enhancement of close to an order of magnitude of the narrow C-H band of chloroform at $\lambda = 3.3\mu$. Figure 5 shows the enhancement of a small portion (at arrow) of the O-H band of

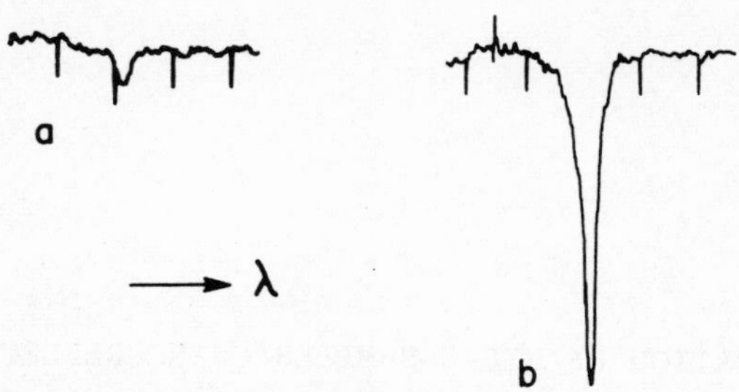

Fig. 4. Comparison of the absorption by the C-H band of chloroform ($\lambda = 3.3\mu$) for ||-polarization at $\theta = 29.5^{o}$ using (a) an uncoated Si prism and (b) an optical cavity designed for ||-polarization. The absorption of 4% in curve (a) is enhanced by over an order of magnitude in curve (b).

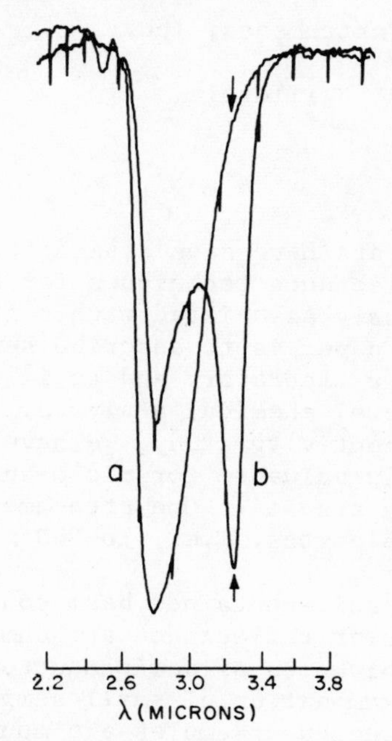

Fig. 5. The O-H absorption band of water at $\theta = 29^{o}$ for ||-polarization using (a) an uncoated Si prism and (b) an optical cavity designed for ||-polarization. The strong amplification for curve (b) at 3.3μ leads to the distortion of the broad O-H band.

water and distortion resulting from the use of a cavity of high Q in the recording of broad bands.

The results and discussion presented here clearly show the promise that Internal Reflection Spectroscopy holds for the detection of minute quantities.

INFRARED EXAMINATION OF MICRO SAMPLES USING REFLECTANCE TECHNIQUES

Kenneth E. Stine

Beckman Instruments, Inc.

Fullerton, California

ABSTRACT

The past few years have seen a significant increase in the use of infrared reflectance techniques for obtaining information of types not previously associated with reflectivity measurements. The purpose of this paper is to describe several reflectance attachments used by our laboratory and to illustrate a few of their applications in general chemical analyses. In addition to the normal types of reflectivity study, we have found these attachments to be extremely valuable for micro-analyses and thin coating or surface chemistry studies. One attachment is particularly useful at extended wavelengths, i.e., to 300 microns.

Because of its self-contained beam condensing system, the use of a micro-specular reflectance attachment, originally designed to measure the thickness of semiconductor epitaxial films, has been used for the examination of small samples, such as those separated from GC effluents. Samples are mounted on a small reflecting surface so as to reflect the light beam through the sample twice, producing a transmission-like spectrum. Due to the double-pass through the sample and the resulting increase in sensitivity, only about one-half the amount of sample normally required for a micro KBr pellet of equivalent area is necessary. In conjunction with gas chromatographic separations, this technique has been used for the identification of DDT in milk and phenobarbital in blood. In favorable cases using a 1x1 mm area of sample, good spectra have been obtained on as little as 15 micrograms of sample without resorting to scale expansion techniques. For more weakly absorbing materials, 25 to 50 micrograms is typical. While the system has been applied mainly to gas chromatographic fractions, it has also been found applicable to the identification of residues from solvent wash systems and LSD from milk sugar.

In those cases where very thin coating must be examined, a multiple specular reflectance attachment has been found to be extremely useful. Basically this system consists of two flat highly reflective surfaces, such as aluminum or steel, held 4 mm apart and oriented at 45° to the infrared spectrophotometer beam. Light entering one end of the attachment is reflected back and forth nine times between the two mirrors before being deflected back to the monochromator via two additional mirrors. Like the microspecular reflectance technique, the multiple reflection system produces transmission-type spectra, but achieves greater sensitivity for a given sample thickness through a multiple reflection process. Whereas a single specular reflection produces spectra about twice as intense as conventional transmission techniques, multiple reflection can produce a spectrum 8, 10 or 18 times more intense, depending on the number of reflections used. Using the multiple specular reflectance and scale expansion techniques, the presence of a typical organic material such as polyvinyl acetate on a highly reflective surface can be detected at the 0.001 micron level.

The growing interest in the far-infrared has prompted the design of several micro beam condensers useful for transmission and reflectance studies in the extended wavelength regions, i.e., beyond 50 microns. Due to the relatively poor efficiency and wavelength limitations imposed by normal beam condensing lenses, front surface mirrors have been used throughout. These systems are typically 80-90% efficient throughout the 2.5-300 micron region and provide a beam size reduction of 4 times in an F-10 system. Use of these systems in an IR-11 far-infrared spectrophotometer allows samples as small as 3x5 mm to be examined at near full instrument energy. Using the condensers in combination with beam aperturing, samples as small as 2 mm^2 in area can be readily examined in the region beyond 50 microns. The results of transmission and reflectance studies in the 33-800 cm^{-1} region are illustrated.

Determination of True Infrared Absorption Frequencies from Internal Reflection Data

E. F. Young, R. W. Hannah

The Perkin-Elmer Corporation

Norwalk, Connecticut 06852

At the 1966 Pittsburgh Conference some measurements of the magnitudes of observed frequency differences between internal reflection bands and transmission bands were presented (1). For strong absorptions with thick films, the internal reflection minima were displaced from the transmission minima toward lower frequencies by as much as 14 cm^{-1}. The reason for these shifts is related, of course, to the large dispersion in the refractive index coefficient and the consequent large changes in reflectivity, which do not coincide with the changes in the absorption coefficient in the vicinity of the absorption bands. Other data in the same paper suggested that, for thin films of the same materials, small frequency shifts occurred in the opposite direction, toward higher frequencies.

These observations quite naturally prompt the rather interesting question--what is the true frequency of the oscillator? Is it the transmission frequency, the high-angle-of-incidence thick-film internal reflection frequency, the thin-film internal reflection frequency, or none of these? Furthermore, this question may be extended to include the effect of window material on the absorption frequency measured by transmission.

Undoubtedly others had worried about these particular problems and a search through the literature revealed that Schatz, Maeda and Kozima (2) had noted that it is important to keep in mind the choice of window material when making optical constant measurements. In addition, Clifford and Crawford (3), using attenuated total reflection data, calculated frequencies which they termed true frequencies of vibrational transitions. They stated that the true frequency was not always the same as the frequency of the maximum absorption. Their calculated values, which differed from the observed frequencies by several reciprocal centimeters toward higher frequencies for strong absorptions, demonstrated this. Differences for weak bands were found to be less and, in some cases, negligible.

In our calculations of true frequencies, we used a slightly different model for the damped oscillator than Clifford and Crawford (3) which produced qualitatively the same results. The very careful evaluation by Dr. Crawford and his group of the optical constants of several materials across strong and weak absorption bands provided a source of reliable optical constant data for our calculations. It is important to note that the optical data presented by Clifford and Crawford (3) was obtained with very carefully designed attenuated total reflection equipment and from equally carefully designed experiments. Their procedures are well documented in the literature and we shall not describe them here. Furthermore, we are not concerned with repeating their work but with the use of their optical constant data to predict magnitudes of frequency shifts in transmission as a function of reflection effects dependent on crystal window material. The following discussion will describe the relation between the natural oscillator frequency and the observed absorption maxima for internal reflection and transmission measurements. In addition, we shall review the effect of polarization on internal reflection measurements and will describe the effect of window material on transmission frequencies.

The frequency at maximum absorption for an infrared absorption band does not, in general, correspond to the natural frequency of the absorbing mechanism, and the band shape does not always reflect the electromechanical coupling of the absorbing mechanism to its environment. Furthermore, the experimental technique of measurement introduces frequency shifts and band distortions which are characteristic of the technique itself.

In order to determine the natural frequency of an oscillator absorbing radiation, the dielectric function of a theoretical liquid having one absorption was represented by the classical dispersion function shown in Equation 1.

$$\epsilon(\nu) = \epsilon_\infty + \frac{S^2}{\nu_o^2 - \nu^2 + i\gamma\nu} \qquad (1)$$

where:

ϵ_∞ = high frequency dielectric constant
S = oscillator strength
ν_o = natural oscillator frequency
γ = damping constant
ν = frequency

To give our theoretical oscillator meaning in terms of a real liquid, the dielectric function was fitted to the optical constant data for the 673 cm^{-1} band for benzene published by Clifford and Crawford. The procedure followed was to choose values for the four parameters ϵ_{∞}, S, ν_0 and γ in the dielectric function, calculate values for the complex refractive index, $N = n-ik = [\epsilon(\nu)]^{\frac{1}{2}}$ where n is the real part of the refractive index and k is the imaginary part, as a function of frequency and then compare these calculated values with Crawford's data until an accurate fit was found. This yielded a natural oscillator frequency of 673 cm^{-1}.

For the internal reflection calculations the fitted dielectric function was substituted into the Fresnel reflection equations for parallel and perpendicular polarized radiation. The results of these calculations are shown in Figure 1. The minima in the curves for both polarizations is very near to 668 cm^{-1}, a 5 cm^{-1} shift from the natural oscillator frequency of 673 cm^{-1}. Weaker bands were found to be shifted less. You will note that the two polarizations behave differently, with the parallel radiation producing a stronger reflection band. For unpolarized radiation, or partially polarized radiation as it occurs in spectrophotometers, particularly grating spectrometers, the experimentally determined band would be a weighted average of the two polarizations. Thus, it is important to realize that the determination of good internal reflection data requires a careful consideration of polarization effects, including instrument polarization.

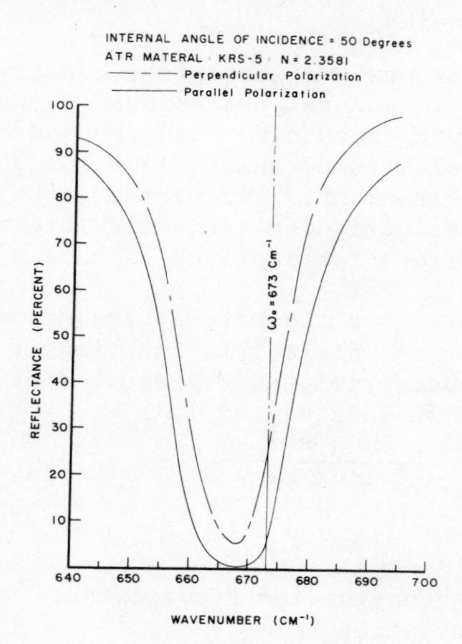

Figure 1
Calculated Internal Reflectance Spectra for the Fitted Dielectric Function

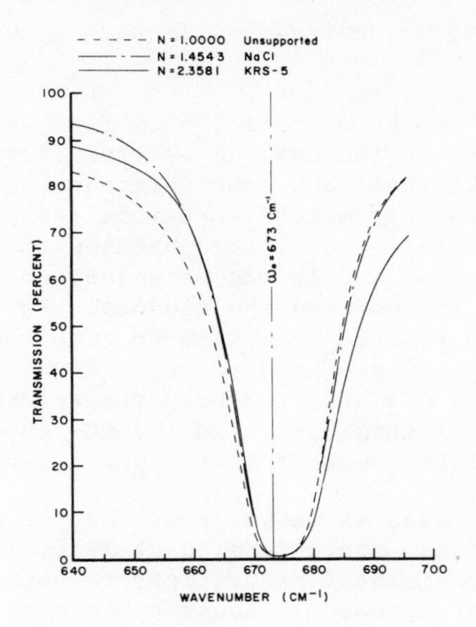

Figure 2
Calculated Transmission Spectra for the Fitted Dielectric Function

The transmission of the theoretical liquid represented by the fitted dielectric function and supported by materials of constant refractive index is shown in Figure 2. From these curves, one can see that the maximum absorption occurs about 3/4 of a wavenumber higher than the natural oscillator frequency at 673 cm^{-1}. In addition, the band shape is quite dependent on the refractive index of the support material. For the high index material, the broadening on the high frequency side of the calculated band tends to cause the band to appear to shift by about 1 cm^{-1} toward higher frequencies as compared with the low index support materials.

Experimental measurements corresponding to the theoretical calculations were made by internal reflection using a single-reflection ATR accessory at 45, 55, and 65° and by transmission using NaCl and KRS-5 windows. Frequencies of absorption maxima for benzene were determined for the strong band centered near 673 cm^{-1} and the weak band near 1035 cm^{-1} as follows. A Perkin-Elmer Model 621 equipped with a frequency marker system set to provide fiducial marks every ½ cm^{-1} was calibrated with carbon dioxide in the 670 cm^{-1} region and with indene in the 1035 cm^{-1} region. Frequency calibration was checked before and after each day's spectra were obtained. Repeatability over the several days required for the measurements was better than 0.15 cm^{-1} and over a given day was better than 0.1 cm^{-1}. In addition, comparison of the mean values of each band used for calibration with values from the International Union of Pure and Applied Chemistry Tables of Wavenumbers (4) indicated the frequencies were accurate to better than 0.5 cm^{-1}.

The benzene absorption near 673 cm^{-1} is very strong, and the location of the absorption maximum by transmission required the use of a demountable cell and some reference beam attenuation. Instrument operating parameters were modified appropriately. Warming of the sample in the instrument led to fairly rapid evaporation of the benzene and it was necessary to continually refill the capillary spacing between the windows. Furthermore, the evaporated benzene contributed from time to time to the absorption pattern, so that purging of the instrument was required. Of course, atmospheric CO_2 affected both the transmission and reflection measurements of this band, and the CO_2 concentration was, therefore, appreciably reduced in the purge gas.

The 1035 cm^{-1} band is weaker than that at 673 cm^{-1}, and 25 μ sealed cells could be used. Purging at 1035 cm^{-1} was not critical. With the experimental precautions taken, the transmission or reflection band was scanned at least ten times, mean values of the absorption maximum calculated and standard deviations determined.

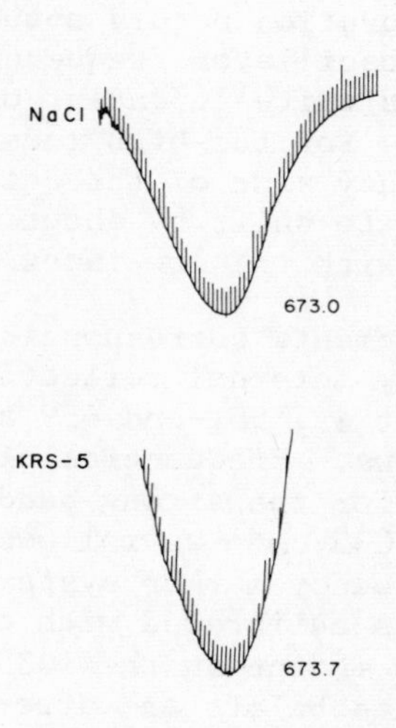

Figure 3
Experimental Transmission Spectra for Benzene

Figure 3 shows examples of the experimental transmission measurements for the 673 cm^{-1} band. Frequency marks are ½ cm^{-1} apart. The band is quite broad and location of the absorption maximum was somewhat difficult. This was reflected in a higher standard deviation among the data. The top curve was made with NaCl windows, the lower with KRS-5. In Table 1 the mean values for the absorption maxima for this band and for the weaker, 1035 cm^{-1} band are summarized. The difference shown for the 673 cm^{-1} absorption between those results labelled NaCl and those labelled KRS-5 is considered significant, but the smaller shift indicated for the 1035 cm^{-1} band, even though in the right direction, corresponds roughly to the experimental error and cannot be considered significant. It is interesting to note that the experimental measurements were done independently of the theoretical calculations, and yet the experimental difference observed for the 673 cm^{-1} band is almost precisely that predicted theoretically.

BENZENE

Window Material	Frequency of Maximum Absorption
NaCl	673.0
KRS-5	673.8
NaCl	1035.8
KRS-5	1036.0

Table 1
Transmission Data for Benzene

Figure 4 shows some examples of the internal reflection spectra for the 673 cm^{-1} band for angles of incidence of 45° and 55° and Table 2 summarizes the experimental data. The differences among the three values are considered significant. In addition, the rather large, 4.4 cm^{-1} difference between even the highest internal reflection frequency of 668.6 cm^{-1} and the transmission value of 673.0 cm^{-1} provides an indication of the magnitude of the refractive index change occurring in the vicinity of the band, since change in this optical property will have the largest effect on the internal reflection value. Once again the magnitude of the shifts and the difference between the reflection and transmission values are consistent with the theoretical predictions.

Early in this discussion the question was raised, "What is the true oscillator frequency?" Unfortunately, we are not yet in a position to provide a final answer. The theoretical calculations can differ by several wavenumbers depending on the damped oscillator model chosen, and the experimental values have been

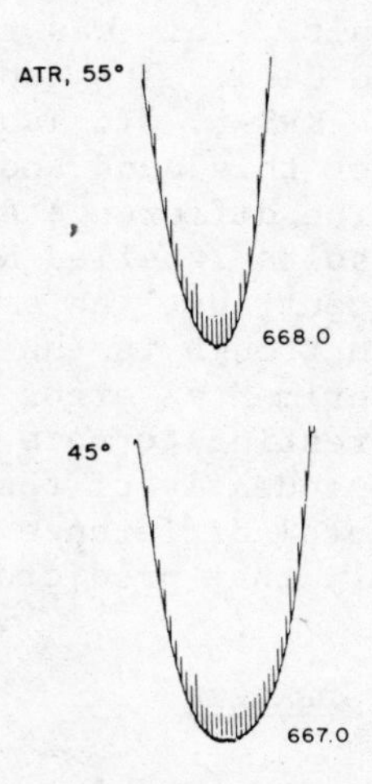

Figure 4
Experimental Internal Reflectance Spectra for Benzene

ATR, 675 cm^{-1} Band of Benzene

Angle of Incidence	Frequency
45°	667.0
55°	668.0
65°	668.6

Table 2
Internal Reflectance Data for Benzene

shown to be dependent on the support material. However, both approaches strongly suggest that some consideration of cell window material must be made in any careful definition of band position, particularly of the stronger, and thus generally, the fundamental absorption. It would seem that this would be especially true in the Coblentz Society specifications for reference spectra (5). Their wavenumber accuracy specification for Class II spectra is ± 3 cm^{-1} below 2000 cm^{-1}, and we have shown that for strong absorptions nearly one third of this error could arise because of an unfortunate choice of window material.

The theoretical plots also showed that the band shape may be broadened toward higher frequencies, depending on the support material. This possible increase in band width must be remembered in any investigation in which conclusions are reached from band shapes and band widths. In addition, the magnitude of the difference between the transmission value and the internal reflection minimum can be considered a qualitative measure of the magnitude of the variation in refractive index across the absorption band. Therefore, the difference is indicative, as well, of the effect of window material on the frequency as measured by transmission. This suggests that perhaps both types of measurements be made on a given sample.

It is worthwhile concluding with the observation that it is not the raw transmission or absorption spectrum which is physically significant but rather the optical constants as a function of frequency, and this data should be considered as reference spectra. When one adds to this the fact that the optical constant data can be extracted reasonably easily from good internal reflection data for liquids, then it becomes obvious that more effort must be directed into the art and science of reflection spectroscopy in the infrared.

1. Robert W. Hannah, Pittsburgh Conference on Analytical Chemistry and Applied Spectroscopy, 1966.

2. P. N. Schatz, Shiro Maeda, Kunio Kozima, J. Chem. Phys. 38, 2658 (1963).

3. A. A. Clifford, Bryce Crawford, Jr., J. Phys. Chem. 70, 1536 (1966).

4. Tables of Wavenumbers for the Calibration of Infra-red Spectrometers, Butterworth Inc., Washington, D. C. (1961).

5. The Coblentz Society Board of Managers, Anal. Chem. 38, 27A (1966).

HIGH RESOLUTION GONIOPHOTOMETER AND ITS USE TO MEASURE APPEARANCE PROPERTIES AND LIGHT-SCATTERING PHENOMENA*

Richard S. Hunter

Hunter Associates Laboratory, Inc.

9529 Lee Highway, Fairfax, Virginia 22030

This paper describes a tool which has been designed primarily for investigating the appearance properties of objects. Specifically, this tool is used to measure and plot automatically those aspects of object appearance associated with the geometric manner in which these objects reflect or transmit light.

This instrument should also be of interest to chemists because of the possibility for use in product identification in addition to its use in appearance measurement.

Goniophotometry is simply the use of an instrument (such as the one diagrammed in Figure 1) to measure the intensity and direction of light reflected and transmitted by materials. For reflection measurements, it is assumed that the specimen is flat. For transmission measurements, it is further assumed that the specimen is in the form of a sheet. A goniophotometer registers curves of the amount of light reflected as functions of the direction of reflection. The goniophotometric curve is therefore related to gloss in the same manner that a spectrophotometric curve is related to color. It is a physical analysis of the properties of the surface responsible for the way the surface looks.

In appearance measurement, goniophotometry is generally limited to the visual luminosity function even though the geometric distribution of light measured with a goniophotometer can, and does,

* Parts of this article are taken from the Chapter on Goniophotometers in the forthcoming Handbook on Color and Appearance Measurement by Richard S. Hunter which is now in preparation.

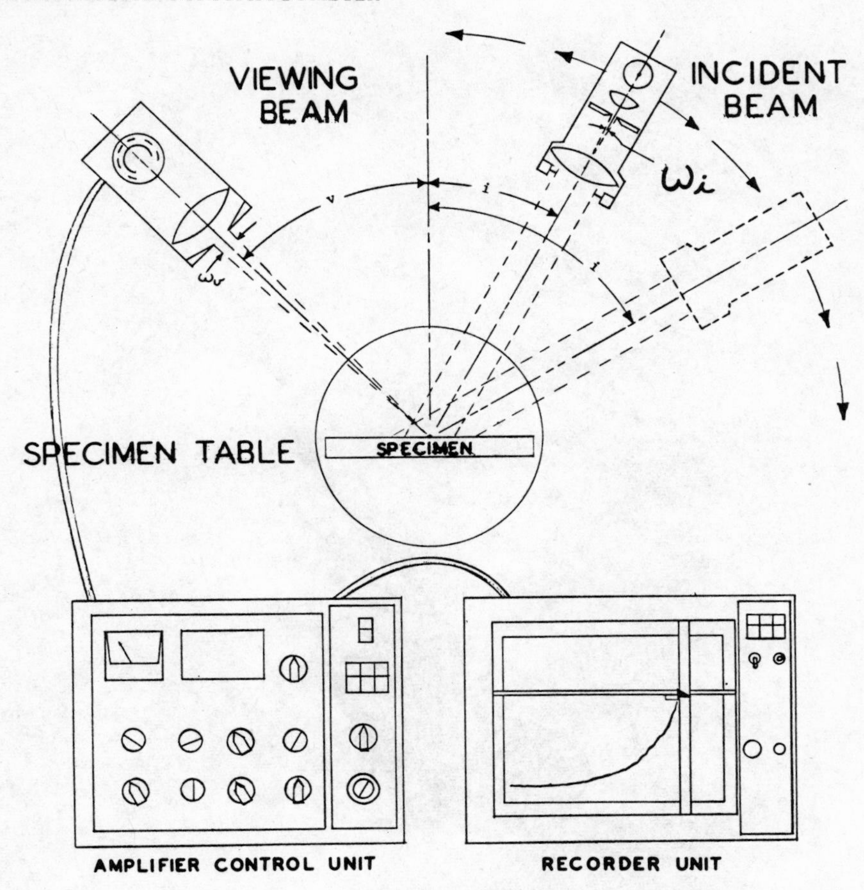

Figure 1. Diagram of the Hunterlab Recording Goniophotometer showing the components of the optical unit above (incident beam, specimen in the center, and viewing beam). Below are the amplifier control unit and the recorder unit on which the continuous curves are drawn.

vary with wavelength. The usual curve is obtained with a filter and a phototube which simulates the CIE $\bar{y}$ response. This visual luminosity response peaks in the yellow-green region of the spectrum.

The chemist, on the other hand, is most frequently interested in measurements at the blue end of the spectrum where the highest turbidities occur. For this reason, and because wavelength is a factor in the particle-size calculation, goniophotometric light scattering measurements for chemical identification are spectrally different and closely controlled as to wavelength.

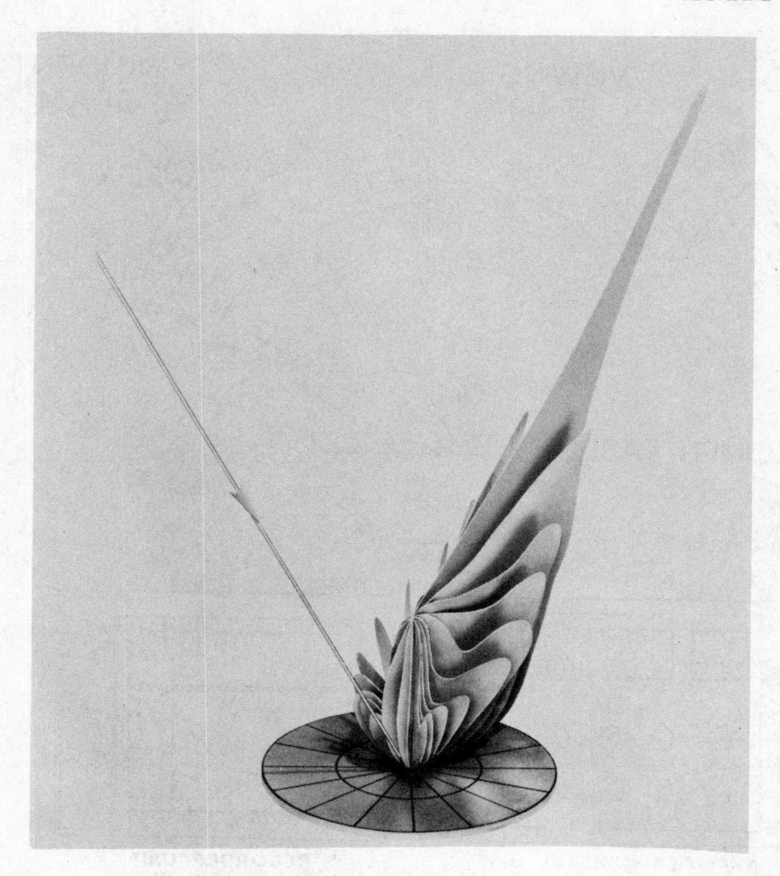

Figure 2. Three-dimensional goniophotometric curve of a sheet of diffuse aluminum (printed by permission of Applied Optics, Vol. 4, 7, p.818 (July 1965).

The reference standard normally used for goniophotometry is the ideal perfect white diffusor; that is, a white which reflects and diffuses completely. Measurements in terms of this standard are called values of directional reflectance. If the actual value at one set of geometric measurements is known for example, 45° illumination and 0° viewing, the instrument will give values of directional reflectance at other angles. Using available techniques, standards can be calibrated on an absolute basis to as close as one or two percent. The same scale of light flux intensity is used for transmission measurements where the results are said to be values of directional transmittance.

Figure 3. Photograph of three panels of translucent plastics in front of a fluorescent lamp bulb.

As illustrations of what goniophotometry does, actual curves can be examined. Figure 2 shows distributions of light reflected by an etched aluminum surface. This figure was published in Applied Optics[1]. Note that measurements in this case were made throughout the hemisphere of directions of view. In this example, linear scales were used. The highest point, of course, represents the peak direction of specular reflection.

The remainder of the goniophotometric data reported in this paper were obtained at Hunterlab and involve only the equatorial directions of illumination and view in the perpendicular plane. Logarithmic scales were used rather than the linear measurement scales used in Figure 2.

Figure 3 is a picture of three sheets of white translucent plastic. Plastic "U" diffuses uniformly in different directions. Plastic "V" obscures the image of the light source but, as can be observed, is not uniformly diffusing. Specimen "W" diffuses quite evenly but it is not opaque enough to completely obscure the image of the light source behind it.

Goniophotometric curves for perpendicular (0°) incidence are plotted in Figure 4 for these same three specimens. Note the break in the curves between the directions of reflection on the left and the directions of transmission on the right. Note the flatness of the "U" curve throughout. Note, also, that "W" curve is relatively flat except at the point where the sharp peak appears to represent the light specularly transmitted through it. The intermediate specimen "V" has a hump-shaped curve which increases toward the direction of specular transmission where it is brightest. The break in the curves between reflectance and transmittance is due to two things:

1. The surfaces were glossy and internal reflection prohibits high grazing angle emittance by either reflection or transmission.

2. The light beam goes off the edges of the specimen at near grazing and the instrument therefore is not accurate at angles near 90°.

Figure 5 shows two reflecting specimens and the curves plotted for 45° incident light on these specimens. The actual specimens are two different waxes on a black enamel automotive paint. They are shown under a desk lamp where the tubular image of the lamp can be seen by the reflection from the specimens. Two things

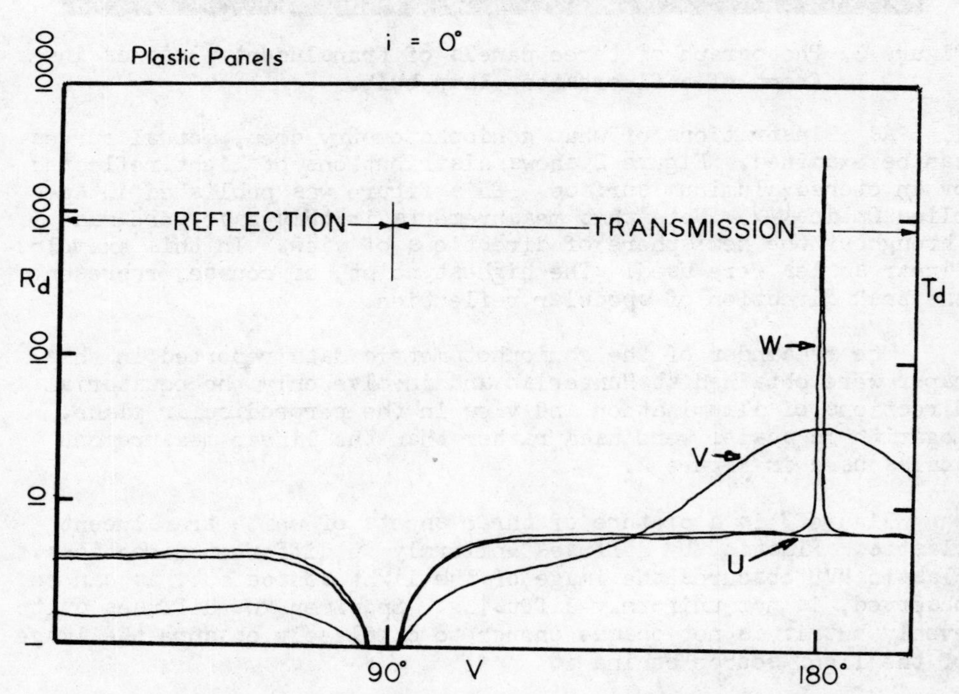

Figure 4. Goniophotometric curves for the three plastic panels shown in Figure 3. Note breaks in curves at point of transition from reflection to transmission. Light was incident perpendicularly (0°) and measurements were taken in directions of both reflection and transmission.

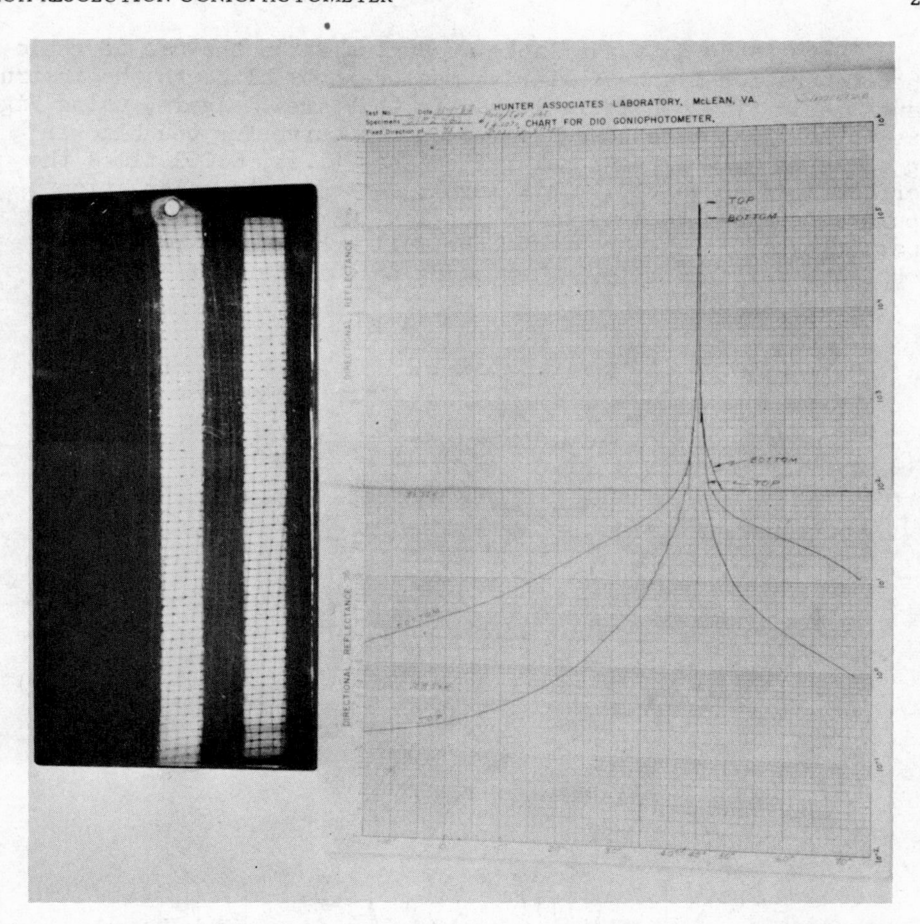

Figure 5. Photograph of goniophotometric curves of two black enamel paints on a panel with curves for the paints beside them. Note that graph shown is pieced together from two four-cycle graph sheets to present eight cycles altogether in one graph. The steepest curve covers six of these eight cycles. The less steep curve represents a hazy sample at the bottom. The cloudiness of the black image area is the visible evidence of this haziness.

are noteworthy about these curves. One is the broad 6-log cycle range of reflection intensity shown, the other is the evidence of haze seen by cloudiness in the dark parts of the mirror images. Note that this haze is higher on the bottom specimen than on the upper. The bottom specimen is, of course, the one whose curve is higher on both sides of the spike representing the specular image peak.

There is no data available at Hunterlab on the use of goniophotometers for light-scattering measurements although the instrument should find considerable use in this area. Again, using Figure 5, note that the lowest value on the curve for the blacker specimen is 0.1% directional reflectance (i.e., 0.001 times the perfect diffuse white). This level, according to calculations, would correspond to a turbidity of 4×10^{-3}, assuming uniform scattering of light throughout the full sphere of directions around the turbid sample. When using this instrument for light

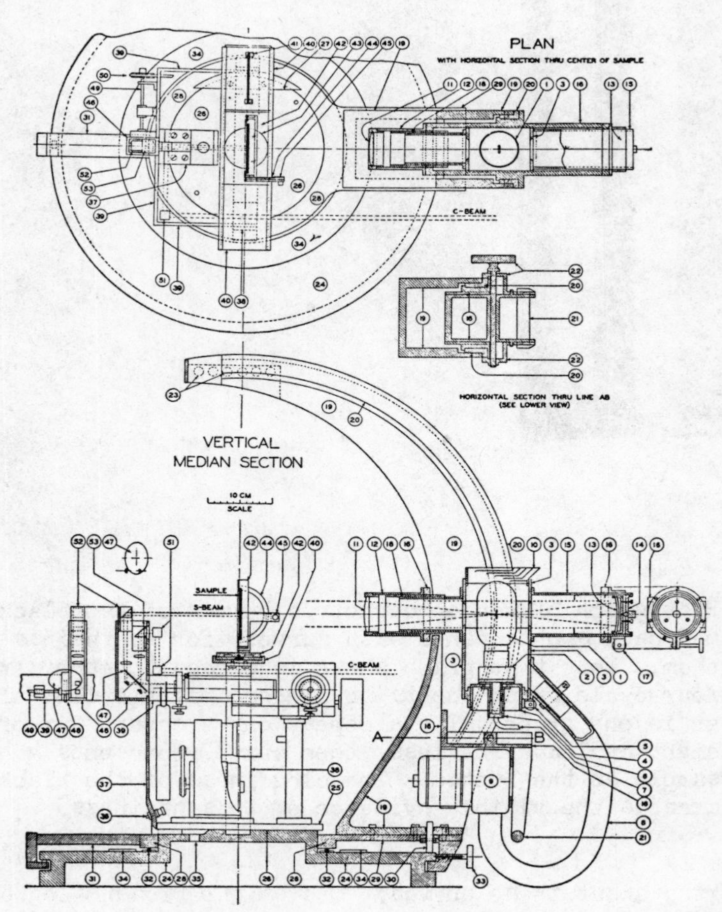

Figure 6. Drawing of a McNicholas Spectrophotometer (See Reference 3). In this instrument the light source is moved around the specimen both horizontally and vertically. The Keuffel & Esser visual spectrophotometer is located directly under the specimen in the bottom center of the diagram.

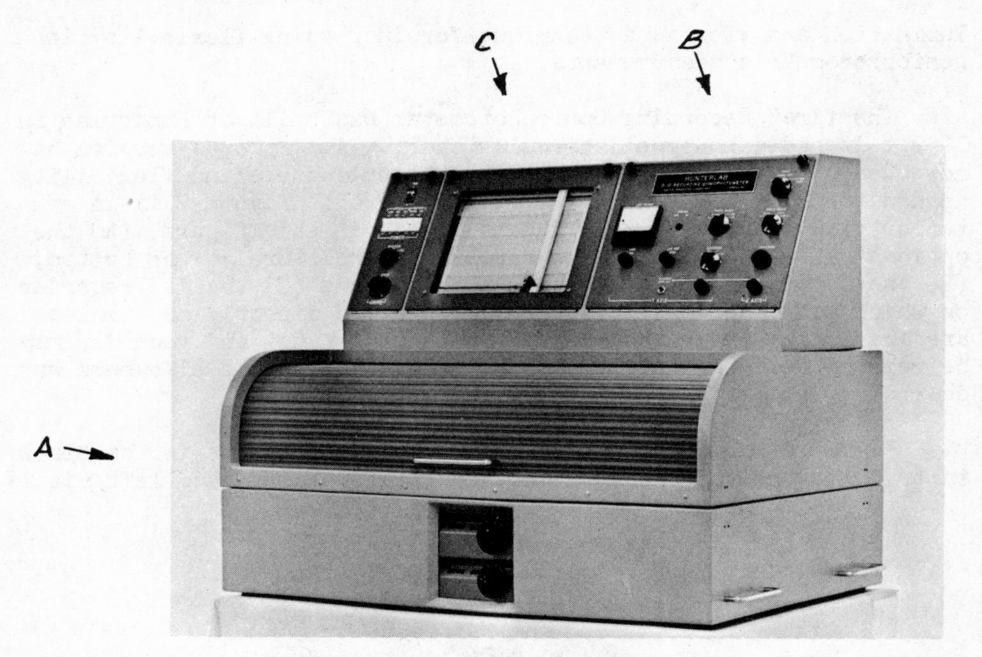

Figure 7. Photograph of Hunterlab D10 Goniophotometer with (A) optical unit at the bottom, (B) control and (C) recorder units above.

scattering studies, it would be wise to baffle and trap against stray light, otherwise the instrument has both the sensitivity and the high angular resolving power needed in this area of effort.

Historically, Jones' evaluation of the reflecting properties of photographic paper in 1922 is probably the first use of goniophotometry[2]. During the 20's and 30's, a number of reports were published on this type of work. Only the work of McNicholas in the early 1930's at the National Bureau of Standards will be mentioned here[3]. Dr. McNicholas could well be called the father of the instrumental methods of reflectance spectroscopy. He built a goniospectrophotometer which is shown in Figure 6. Note that this instrument consists of a spectrophotometer below and a goniophotometer above. The instrument provides a Z-axis for scanning specimens. The only disadvantage of this particular instrument is that measurements were very time-consuming since it is based on visual techniques for measurement of reflected light. Weeks could be spent measuring a single sample. Dr. McNicholas introduced in another paper[4] the concept of a reflectance scale using the perfect white diffusor as a reference standard. In this paper, he also called attention to the usefulness of the Helmholz Reciprocal Relation. This relation justifies the interchange of directions of illumination and view as a technique for increasing flexibility in goniophotometric measurements.

The first Recording Goniophotometer was built at Hunterlab in 1953. It had a Honeywell circular chart. An improved version has now been produced at Hunterlab with the ruggedness and flexibility needed for industrial and university research. Figure 7 is an exterior view of this new instrument. Note in this figure, (A) the optical unit which is under the roll-top enclosure at the bottom, (B) the control panel on the right above, and (C) the X/Y recorder on which goniophotometric curves are plotted directly as samples are scanned by the instrument. The control panel and recorder may be removed from the top of the instrument and placed elsewhere when desired.

Figure 8 shows details of the optical components in the black interior of the instrument. Note the fixed beam on the left ; it is

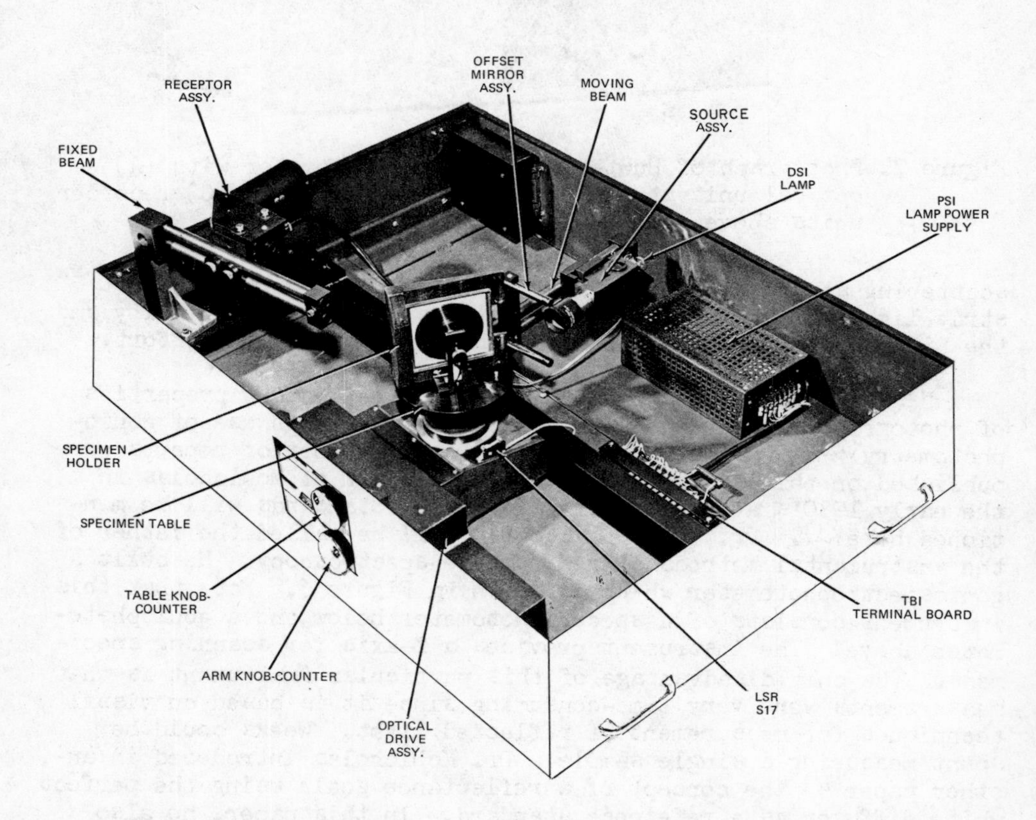

Figure 8. Photograph of optical unit of Goniophotometer showing positions of parts.

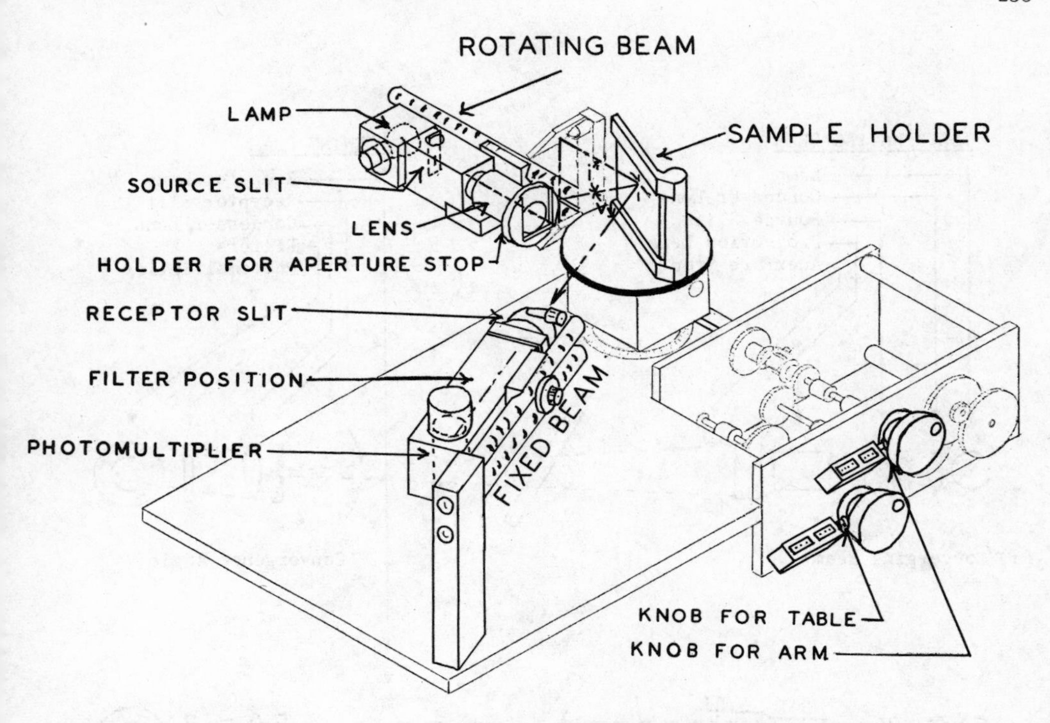

Figure 9. Drawing of optical unit of Goniophotometer showing positions of the two beams, the holder for the samples, and the knobs and dials which control angular positions of the light beams and give these angles to two-tenths of degree.

usually the light receiver. The movable beam is behind and somewhat to the right. It swings through an arc of more than 180°. It is driven by a small gear motor. The specimen holder is above the central table.

Figure 9 is a drawing of the optical unit. The directions of incidence and view may be read out directly on counters to the nearest 1/10th degree. Both table angle with respect to fixed beam and the starting point for scanning may be adjusted by the operator from the instrument exterior.

The aperture stop is selected to give the beam the desired size and shape. The maximum diameter is about 3/4 x 1" for an eliptically shaped beam. Source and receiver slits are selected as needed to adjust the conflicting needs of high angular resolving power and adequate light signal for measurement.

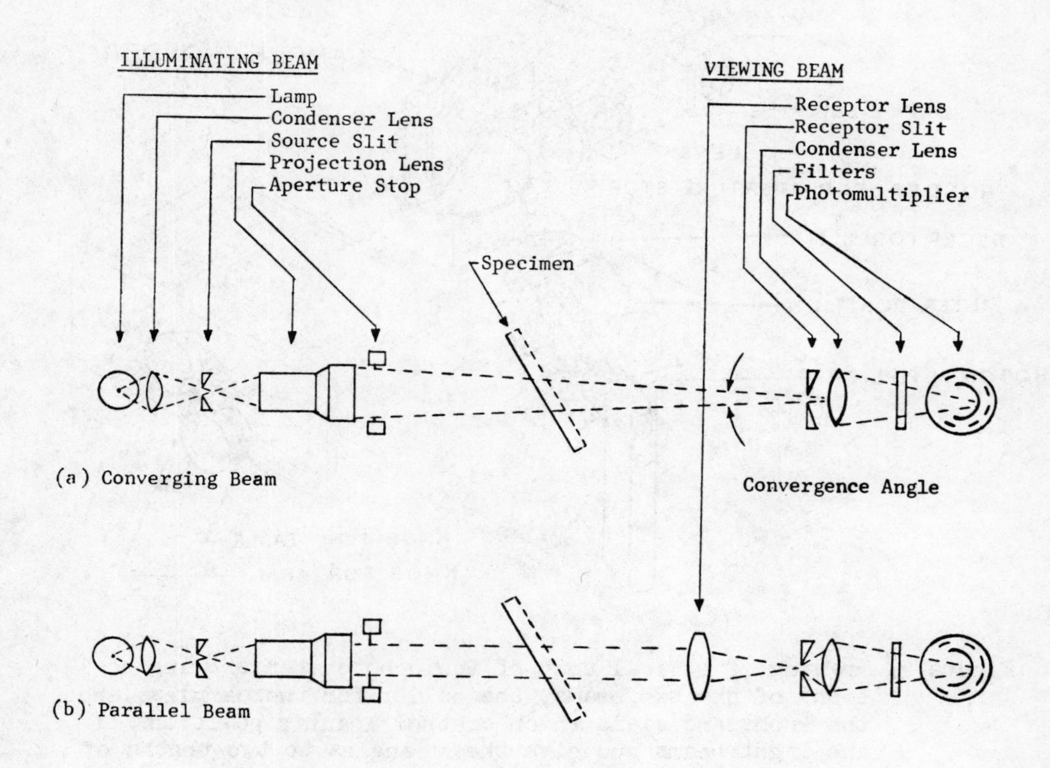

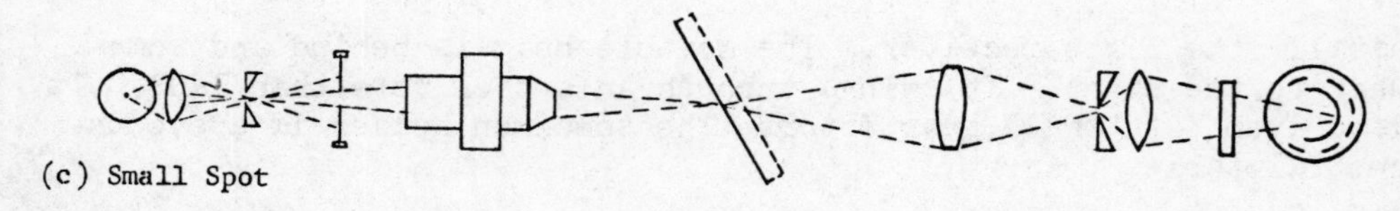

Figure 10. Diagrams of the three beam arrangements which can be used in the spectrophotometer. The converging beam arrangement is simplest and most popular; but the parallel beam arrangement gives true columnation and must be used in some cases. The small spot arrangement is used for single filaments of yarn or for very small specimen areas such as one encounters within a single letter of printed page.

The instrument uses a #6199, 1-1/2" photomultiplier phototube with a Sweet type logarithmic amplifier[5]. This responds over four log cycles, but one-cycle and two-cycle chart papers and ranges are available for use with specimens which diffuse light more uniformly.

Normally, both table and fixed beam remain stationary while the rotating beam scans. Scan ranges of 180° total, 90°, and 18° are regularly used. A 1.8° chart-scan range was prepared especially for some studies of high-gloss metals. Hunterlab has also built into the instrument a capability to turn the table at 1/2 the scan arm rate of movement so that changes of specular reflectance with specular angle can be measured.

Figure 10 shows the three different optical beam arrangements which have been used by Hunterlab. The first is the converging beam arrangement in which a single projection lens forms a specular image of the source slit at the receiver slit. This is the simplest and best arrangement for high resolution work. All the measurements demonstrated in this article were made with the converging beam arrangement. Even though the principal rays at the specimen are not parallel, the arrangement nevertheless gives precise control of the departure of reflected rays from the specular angle. This difference from specular angle is normally the angle of primary interest in goniophotometry.

The second arrangement as shown in Figure 10 is the parallel beam arrangement in which all rays incident on the specimen are essentially parallel. This arrangement presents problems of beam alignment and beam clearance of the second lens when the two beams are near overlap. With the converging beam arrangement, within 8° of overlap can be attained and complete overlap by the use of a beam splitter is planned.

The third beam (Figure 10) is for very small test areas. With this arrangement,areas as small as 0.005" have been used. This is about 0.1 of a millimeter. With this small specimen spot arrangement, angular resolution of specular images is poor. It is limited by the width of the aperture stop measured from the small spot, normally between 2 and 10 degrees.

In the appearance test area, the applications of this instrument are divided into four general types of measurements.

1. Analysis of geometric distribution of reflected light for fixed angle of incidence (see Figs. 2, 4, 5, 11, 12 and 14).

2. Analysis of geometric distribution of transmitted light (see Fig. 4).

3. Analysis of specular reflectance as a function of specular angle (see Fig. 13). In this instance, the specimen turns at 1/2 the rate of the viewing beam and is used for studies of surface structure and determinations of refractive index by Brewster's angle.

4. Expanded details of specular peaks, both reflectance and transmittance (see Fig. 14).

Figures which follow show examples of each of these applications.

Figure 11 compares the diffusion of a flat white paint with a surface which has been called a diffuse metal; specifically an etched aluminum. Notice how flat the curve is for paint by comparison with that for a diffuse metal. Another type of distribution study is shown in Figure 12. Here is a study of distributions of light by two newsprint paper samples for light at -75° incidence. This study was accomplished to determine the best angular conditions for gloss measurements of newsprint samples.

Figure 13 illustrates the third type of so-called specular-angle analysis. Here is a comparison of black glass with two different papers. Paper "C" was shiny but wavy. Paper "B" was less shiny at the near perpendicular angle but, structurally, it was smooth so that its specular angle reflectance increased more rapidly at the higher angles of reflection.

Figure 14 shows expanded detail of specular peaks. Here are comparisons of reflectance curve peaks for a number of different metal foils. Incidentally, the curve for a perfectly polished mirror is 0.07° wide at half peak value when the present instrument is operating with its highest resolution.

Summary. The Recording Goniophotometer is a multipurpose tool which should enjoy much wider use in the future than in the past. This wider use waits only for broader understanding of its capabilities.

A goniophotometer analyzes the geometry of light reflection and transmission by objects. This geometry varies with surface structure, internal structure, and the optical constants of the ingredients.

For the measurement of object appearance, the goniophotometer is analogous to the spectrophotometer. It measures geometric properties in the same way the spectrophotometer measures spectral properties. It can be used to study commercially important examples of differences in gloss (such as are shown in Figs. 5, 12 and 14), or in translucency (see Fig. 4). From these analyses, it is

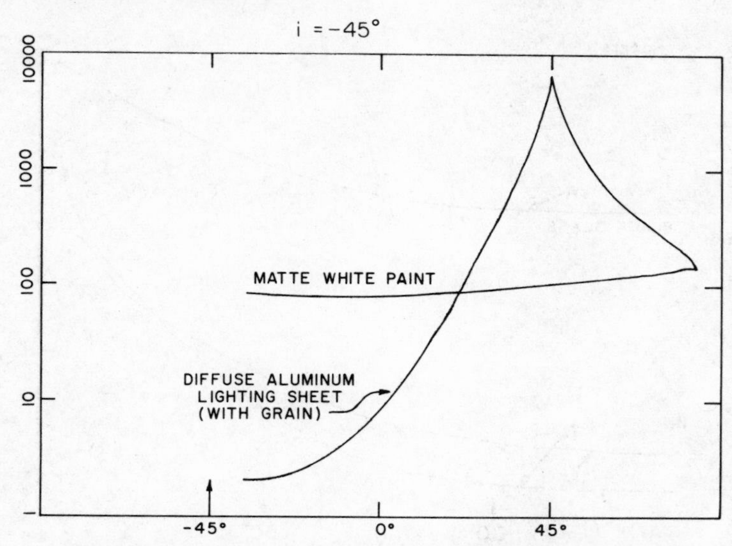

Figure 11. Goniophotometric curves comparing the distribution of reflected light by a white paint (non-metallic) and by a sheet of diffuse aluminum lighting sheet (metallic).

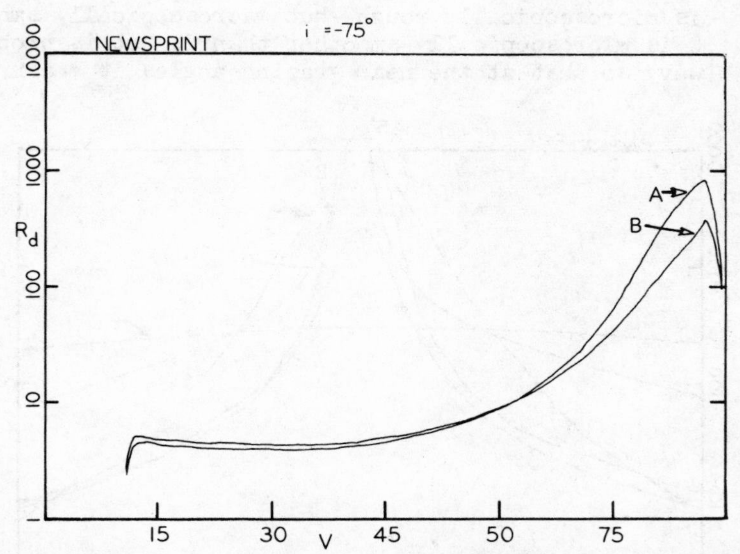

Figure 12. Goniophotometric curves for -75° incidence of two pieces of newsprint. These measurements were made to determine best angles for contrast gloss measurements. Sample A is the glossiest. Note that the curves cross at 55°, thus in setting up a contrast method, one would use as specular angle 75° or even higher. The diffuse angle should be small - 45° or less.

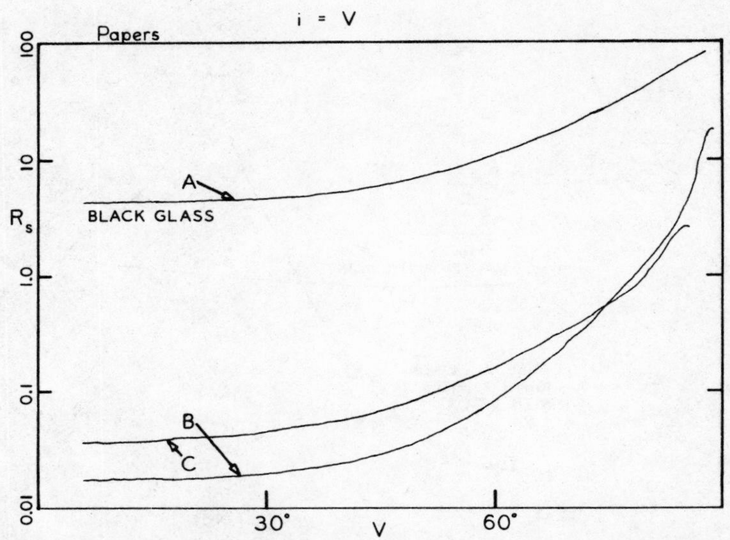

Figure 13. Curves of change of specular reflectance with angle for black glass and for two sheets of paper. Note that curve B is less shiny at low angles, but is shinier than C at angles above 75°. This means that surface B is microscopically rough but macroscopically smooth. C is microscopically smoother than B, but is probably wavy so that at the near grazing angles it measures low.

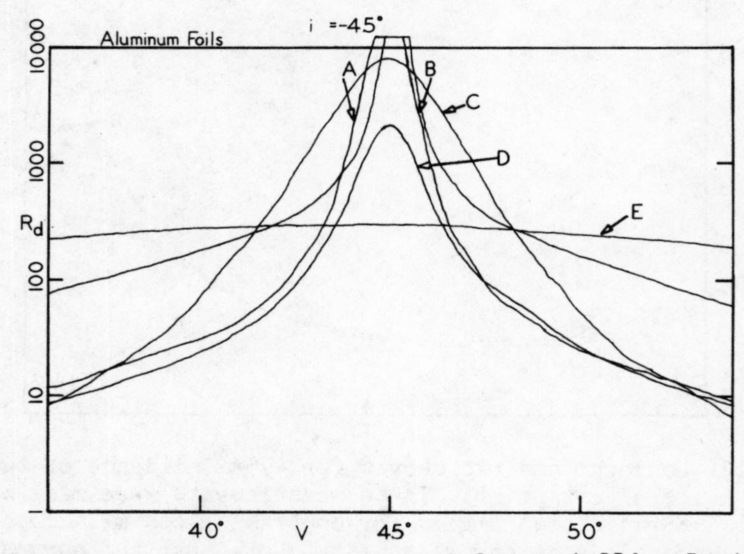

Figure 14. Detail in the curve peak of four metallic aluminum foils compared with reflection from a matte white paint (Specimen E).

possible to identify the different fixed-angle conditions of measurement useful for gloss measurement of specific products. When these fixed-angle directions are identified, they can be incorporated into simpler glossmeters and transmission photometers for routine measurements.

REFERENCES

1. F. E. Nicodemus, Directional Reflectance and Emissivity of an Opaque Surface, Vol. 4, 7, p. 818 (July 1965).
2. L. A. Jones, The Gloss Characteristics of Photographic Papers, J. Opt. Soc. Am. 6, 140 (1922).
3. H. J. McNicholas, Equipment for Measuring the Reflective and Transmissive Properties of Diffusing Media, J. Research NBS 13, 211 (1934) RP704.
4. H. J. McNicholas, Absolute Methods in Reflectometry, BS J. Research 1, 29 (1928) RP3.
5. M. H. Sweet, An Improved Photomultiplier Tube Color Densitometer, Jour. SMPTE 54, 35 (Jan. 1950).

A NEW METHOD OF SAMPLE PREPARATION FOR INFRARED ATR SPECTROPHOTOMETRY

Stanley E. Polchlopek, William J. Menosky, and
Louis G. Dalli

Barnes Engineering Company

Stamford, Connecticut 06902

Pyrolysis has proved to be a useful and interesting adjunct to the ATR technique. Samples containing fillers such as carbon can be pyrolyzed under various conditions of time and temperatures and the pyrolyzates deposited directly on an ATR crystal. This procedure enables the analyst to by pass an extraction or other separation step. A further advance in this technique has been made by putting a high voltage charge on the crystal in order to facilitate the deposit of pyrolyzates on the crystal. When used in this manner the crystal functions as one side of a capacitor. This technique makes it possible to use smaller samples and in some cases lower temperatures. Applications based on the pyrolysis of various small molecules as well as polymers will be discussed. The apparatus used to prepare the condensates on the ATR crystal will be described. Similarities to and departures from the spectra of the parent molecules when compared to the spectra of pyrolyzates will be considered.

INFRARED SPECTRAL REFLECTANCE INDEPENDENT OF EMITTED FLUX AT CONTROLLED TEMPERATURES

Harry G. Keegan and Victor R. Weidner

Clemson University, Clemson, South Carolina 29631 and

National Bureau of Standards, Washington, D. C. 20234

This research program consisted of the development of techniques and instrumentation for obtaining infrared spectral reflectance measurements independent of emitted flux at controlled temperatures for spectral range 4000 to 450 wavenumbers (2.5 to 22.2 microns) at temperatures ranging from -175°C (liquid Nitrogen temperatures) to approximately +1000°C. Supported by ARPA, a Cary-White infrared spectrophotometer was used equipped with a special designed (White, J. Opt. Soc. Am. 54, 1332, 1964) hemispherical reflectance attachment with temperature controls of heated substrates, water cooled sample holder, nichrome source input power control, heated sample holder, and specially designed liquid nitrogen cell holder. Measurements were made of frosts of H_2O and CO_2 (Keegan and Weidner, J. Opt. Soc. Am. 56, 523, 1966) N_2O_4, NH_3, combinations of them, and dry ice; natural and man-made objects; highly absorbing materials used for black-body cavities; highly reflecting metals; the effect of temperature on painted surfaces, and time of exposure; and the effect of temperature on some rare-earth oxides. From spectral reflectance data independent of emitted flux at controlled temperatures, emissivity data can be obtained indirectly on opaque materials from the relationship: emissivity is equal to one minus the reflectance. This is particularly useful in obtaining emissivity data for samples at temperatures below 200°C where direct measurements of emissivity are not easily obtained on ablative materials such as polytetrafluoroethelyene.

PRECISION AND ACCURACY OF SPECTROPHOTOMETRIC REFLECTANCE MEASUREMENTS

Joseph L. Rood and J. R. Hensler

Bausch and Lomb, Inc.

Rochester, New York 14602

The Inter-Society Color Council has made a study (1) of the reproducibility within a given laboratory of the General Electric recording spectrophotometer, and the variability of different instruments from one laboratory to another. The samples used as standards were ten Carrara glass plaques from Pittsburgh Plate and four colored acrylic plaques from Du Pont. A different study (2) compared a General Electric instrument and a Bausch & Lomb Spectronic 505 within the same laboratory. The consitions under which the studies were carried out will be discussed, as will the results. Both studies were concerned with transmittance as well as reflectance measurements. Only reflectance will be discussed, except to mention that in all cases the variations in reflectance measurements exceeded those for transmittance.

References

1. F. W. Billmeyer, J. Opt. Soc. Amer 55, 694 (1965).

2. J. L. Rood, Die Farbe, Heft 1-4, 105 (1961).

INFRARED INVESTIGATIONS OF THE STRUCTURE OF NON-OXIDE CHALCOGENIDE GLASSES

A. Ray Hilton and Charlie E. Jones

Texas Instruments, Inc.

Dallas, Texas 75222

The structural nature of non-oxide chalcogenide glasses was investigated using infrared reflection and absorption techniques. The wave number corresponding to stretching mode vibrations for nine pairs of metal-chalcogen constituent atoms was identified using reflection and absorption spectra obtained from glasses representing fifteen chalcogenide glass systems containing IVA and/or VA elements. The validity of the vibrational assignments were verified by assuming them to represent simple diatomic stretching modes, applying Gordy's simple diatomic force constant relation, and calculating the equilibrium interatomic distance for each bond pair. Comparison of calculated values with the sum of the appropriate covalent radii showed reasonable agreement in all cases, except for the As-S and As-Se vibrations, indicating the local symmetry for glasses containing these elements was different from the others. Using Somayajulu's rule for calculating multi-bond force constants, the normal vibrational frequencies for nine constituent atom pairs were calculated for a $X-Y_2$ linear symmetric triatomic configuration, a $X-Y_2$ non-linear symmetric triatomic configuration, a $X-Y_3$ pyramidal arrangement and a $X-Y_4$ tetrahedral arrangement. Comparison between calculated and observed frequencies indicate that in most cases the IVA element forms a bridge between chalcogen atoms forming a $X-Y_2$ non-linear symmetric triatomic configuration, while As in S and Se glasses exists in a pyramidal $X-Y_3$ configuration.

AUTHOR INDEX

SUBJECT INDEX